FAMOUS PROBLEMS OF ELEMENTARY GEOMETRY

THE DUPLICATION OF THE CUBE
THE TRISECTION OF AN ANGLE
THE QUADRATURE OF THE CIRCLE

An Authorized Translation of F. Klein's
Vorträge Über Ausgewählte Fragen der Elementargeometrie
Ausgearbeitet Von F. Tägert

BY

WOOSTER WOODRUFF BEMAN

AND

DAVID EUGENE SMITH

Second Edition Revised, and Enlarged with Notes by

RAYMOND CLARE ARCHIBALD

DOVER PUBLICATIONS, INC.
Mineola, New York

DOVER PHOENIX EDITIONS

Bibliographical Note

This new Dover edition, first published in 2003, is an unabridged and unaltered republication of the 1956 Dover reprint of the 1930 second revised edition which was enlarged with notes by Raymond Clare Archibald.

Library of Congress Cataloging-in-Publication Data

Klein, Felix, 1849-1925.
[Vorträge über ausgewahlte Fragen der Elementargeometrie. English]
Famous problems of elementary geometry : the duplication of the cube, the trisection of an angle, the quadrature of the circle / by Wooster Woodruff Beman and David Eugene Smith.
p. cm.—(Dover phoenix editions)
Previously published: 2nd and rev. ed., and enlarged with notes. New York : Dover Publications, 1956.
"An Authorized translation of F. Klein's Vorträge über ausgewahlte Fragen der Elementargeometrie, Ausgearbeitet Von F. Tägert."
ISBN 0-486-49551-5
1. Geometry—Problems, Famous. I. Beman, Wooster Woodruff, 1850-1922. II. Smith, David Eugene, 1941- III. Title. IV. Series.

QA466.K5413 2003
516.2'04—dc22

2003055584

Manufactured in the United States of America
Dover Publications, Inc., 31 East 2nd Street, Mineola, N.Y. 11501

PREFACE.

THE more precise definitions and more rigorous methods of demonstration developed by modern mathematics are looked upon by the mass of gymnasium professors as abstruse and excessively abstract, and accordingly as of importance only for the small circle of specialists. With a view to counteracting this tendency it gave me pleasure to set forth last summer in a brief course of lectures before a larger audience than usual what modern science has to say regarding the possibility of elementary geometric constructions. Some time before, I had had occasion to present a sketch of these lectures in an Easter vacation course at Göttingen. The audience seemed to take great interest in them, and this impression has been confirmed by the experience of the summer semester. I venture therefore to present a short exposition of my lectures to the Association for the Advancement of the Teaching of Mathematics and the Natural Sciences, for the meeting to be held at Göttingen. This exposition has been prepared by Oberlehrer Tägert, of Ems, who attended the vacation course just mentioned. He also had at his disposal the lecture notes written out under my supervision by several of my summer semester students. I hope that this unpretending little book may contribute to promote the useful work of the association.

F. KLEIN.

GÖTTINGEN, Easter, 1895.

TRANSLATORS' PREFACE.

At the Göttingen meeting of the German Association for the Advancement of the Teaching of Mathematics and the Natural Sciences, Professor Felix Klein presented a discussion of the three famous geometric problems of antiquity, — the duplication of the cube, the trisection of an angle, and the quadrature of the circle, as viewed in the light of modern research.

This was done with the avowed purpose of bringing the study of mathematics in the university into closer touch with the work of the gymnasium. That Professor Klein is likely to succeed in this effort is shown by the favorable reception accorded his lectures by the association, the uniform commendation of the educational journals, and the fact that translations into French and Italian have already appeared.

The treatment of the subject is elementary, not even a knowledge of the differential and integral calculus being required. Among the questions answered are such as these: Under what circumstances is a geometric construction possible? By what means can it be effected? What are transcendental numbers? How can we prove that e and π are transcendental?

With the belief that an English presentation of so important a work would appeal to many unable to read the original,

Professor Klein's consent to a translation was sought and readily secured.

In its preparation the authors have also made free use of the French translation by Professor J. Griess, of Algiers, following its modifications where it seemed advisable.

They desire further to thank Professor Ziwet for assistance in improving the translation and in reading the proof-sheets.

W. W. BEMAN.
D. E. SMITH.

August, 1897.

EDITOR'S PREFACE.

Within three years of its publication thirty-five years ago Klein's little work was translated into English, French, Italian, and Russian[1]. In the United States it filled a decided need for many years, and not a few teachers regretted that the work was allowed to go out of print. No other work supplied in such compact form just the information here found. Hence it seemed desirable to have a new edition with at least some of the slips of the first edition rectified, and with added notes illuminating the text.

The corrections and notes of the present edition are little more than revised extracts from my article in *The American Mathematical Monthly*[2], 1914. I am indebted to the Editors for courteously allowing the reproduction of this material.

R. C. A.

February, 1930.

---

[1] French translation by Griess, Paris, Nony, 1896; Italian by Giudice, Turin, Rosenberg e Sallier, 1896; Russian by Parfentiev and Sintsov, Kazan, 1898. This last translation seems to have been unknown to the editors of Klein's *Abhandlungen* (see v. 3, 1923, p. *28*).

[2] Remarks on Klein's "Famous Problems of Elementary Geometry", v. 21, p. 247—259.

CONTENTS.

INTRODUCTION.

PART I.

The Possibility of the Construction of Algebraic Expressions.

Chapter I. Algebraic Equations Solvable by Square Roots.

Chapter II. The Delian Problem and the Trisection of the Angle.

Chapter III. The Division of the Circle into Equal Parts.

CHAPTER IV. THE CONSTRUCTION OF THE REGULAR POLYGON OF 17 SIDES.

CHAPTER V. GENERAL CONSIDERATIONS ON ALGEBRAIC CONSTRUCTIONS.

PART II.

Transcendental Numbers and the Quadrature of the Circle.

CHAPTER I. CANTOR'S DEMONSTRATION OF THE EXISTENCE OF TRANSCENDENTAL NUMBERS.

CHAPTER II. HISTORICAL SURVEY OF THE ATTEMPTS AT THE COMPUTATION AND CONSTRUCTION OF π.

CHAPTER III. THE TRANSCENDENCE OF THE NUMBER e.

CHAPTER IV. THE TRANSCENDENCE OF THE NUMBER π.

CHAPTER V. THE INTEGRAPH AND THE GEOMETRIC CONSTRUCTION OF π.

INTRODUCTION.

THIS course of lectures is due to the desire on my part to bring the study of mathematics in the university into closer touch with the needs of the secondary schools. Still it is not intended for beginners, since the matters under discussion are treated from a higher standpoint than that of the schools. On the other hand, it presupposes but little preliminary work, only the elements of analysis being required, as, for example, in the development of the exponential function into a series.

We propose to treat of geometrical constructions, and our object will not be so much to find the solution suited to each case as to determine the *possibility* or *impossibility* of a solution.

Three problems, the object of much research in ancient times, will prove to be of special interest. They are

1. *The problem of the duplication of the cube* (also called the *Delian problem*).
2. *The trisection of an arbitrary angle.*
3. *The quadrature of the circle, i.e.*, the construction of π.

In all these problems the ancients sought in vain for a solution with straight edge and compasses, and the celebrity of these problems is due chiefly to the fact that their solution seemed to demand the use of appliances of a higher order. In fact, we propose to show that a solution by the use of straight edge and compasses is impossible.

The impossibility of the solution of the third problem was demonstrated only very recently. That of the first and second is implicitly involved in the Galois theory as presented to-day in treatises on higher algebra. On the other hand, we find no explicit demonstration in elementary form unless it be in Petersen's text-books, works which are also noteworthy in other respects.

At the outset we must insist upon the difference between *practical* and *theoretical* constructions. For example, if we need a divided circle as a measuring instrument, we construct it simply on trial. Theoretically, in earlier times, it was possible (*i.e.*, by the use of straight edge and compasses) only to divide the circle into a number of parts represented by 2^n, 3, and 5, and their products. Gauss added other cases by showing the possibility of the division into parts where p is a prime number of the form $p = 2^{2^\mu} + 1$, and the impossibility for all other numbers. No practical advantage is derived from these results; *the significance of Gauss's developments is purely theoretical.* The same is true of all the discussions of the present course.

Our fundamental problem may be stated: *What geometrical constructions are, and what are not, theoretically possible?* To define sharply the meaning of the word "construction," we must designate the instruments which we propose to use in each case. We shall consider

1. Straight edge and compasses,
2. Compasses alone,
3. Straight edge alone,
4. Other instruments used in connection with straight edge and compasses.

The singular thing is that elementary geometry furnishes no answer to the question. We must fall back upon algebra and the higher analysis. The question then arises: How shall we use the language of these sciences to express the employment of straight edge and compasses? This new method of attack is rendered necessary because elementary geometry possesses no general method, no *algorithm*, as do the last two sciences.

In analysis we have first *rational* operations: addition, subtraction, multiplication, and division. These operations can be directly effected geometrically upon two given segments by the aid of proportions, if, in the case of multiplication and division, we introduce an auxiliary unit-segment.

Further, there are *irrational* operations, subdivided into *algebraic* and *transcendental*. The simplest algebraic operations are the extraction of square and higher roots, and the solution of algebraic equations not solvable by radicals, such as those of the fifth and higher degrees. As we know how to construct $\sqrt{ab}$, rational operations in general, and irrational operations involving only square roots, can be constructed. On the other hand, every *individual* geometrical construction which can be reduced to the intersection of two straight lines, a straight line and a circle, or two circles, is equivalent to a rational operation or the extraction of a square root. In the higher irrational operations the construction is therefore impossible, *unless we can find a way of effecting it by the aid of square roots.* In all these constructions it is obvious that the number of operations must be limited.

We may therefore state the following fundamental theorem: *The necessary and sufficient condition that an analytic expression can be constructed with straight edge and compasses is that it can be derived from the known quantities by a finite number of rational operations and square roots.*

Accordingly, if we wish to show that a quantity cannot be constructed with straight edge and compasses, we must prove that the corresponding equation is not solvable by a finite number of square roots.

A fortiori the solution is impossible when the problem has *no* corresponding algebraic equation. An expression which satisfies no algebraic equation is called a transcendental number. This case occurs, as we shall show, with the number π.

PART I

THE POSSIBILITY OF THE CONSTRUCTION OF ALGEBRAIC EXPRESSIONS

PART I

THE POSSIBILITY OF THE CONSTRUCTION OF ALGEBRAIC EXPRESSIONS

CHAPTER I.

Algebraic Equations Solvable by Square Roots.

The following propositions taken from the theory of algebraic equations are probably known to the reader, yet to secure greater clearness of view we shall give brief demonstrations.

If x, *the quantity to be constructed, depends only upon rational expressions and square roots, it is a root of an irreducible equation* $\phi(x)=0$, *whose degree is always a power of* 2.

1. To get a clear idea of the structure of the quantity x, suppose it, *e.g.*, of the form

$$x=\frac{\sqrt{a+\sqrt{c+ef}}+\sqrt{d+\sqrt{b}}}{\sqrt{a}+\sqrt{b}}+\frac{p+\sqrt{q}}{\sqrt{r}},$$

where a, b, c, d, e, f, p, q, r are rational expressions.

2. The number of radicals one over another occurring in any term of x is called the *order of the term;* the preceding expression contains terms of orders 0, 1, 2.

3. Let μ designate the *maximum order*, so that no term can have more than μ radicals one over another.

4. In the example $x = \sqrt{2} + \sqrt{3} + \sqrt{6}$, we have three expressions of the first order, but as it may be written

$$x = \sqrt{2} + \sqrt{3} + \sqrt{2} \cdot \sqrt{3},$$

it really depends on only two distinct expressions.

We shall suppose that this reduction has been made in all the terms of x, *so that among the* n *terms of order* μ *none can be expressed rationally as a function of any other terms of order* μ *or of lower order.*

We shall make the same supposition regarding terms of the order $\mu - 1$ or of lower order, whether these occur explicitly or implicitly. This hypothesis is obviously a very natural one and of great importance in later discussions.

5. Normal Form of x.

If the expression x is a sum of terms with different denominators we may reduce them to the same denominator and thus obtain x as the quotient of two integral functions.

Suppose $\sqrt{Q}$ one of the terms of x of order μ; it can occur in x only explicitly, since μ is the maximum order. Since, further, the powers of $\sqrt{Q}$ may be expressed as functions of $\sqrt{Q}$ and Q, which is a term of lower order, we may put

$$x = \frac{a + b\sqrt{Q}}{c + d\sqrt{Q}},$$

where a, b, c, d contain no more than $n - 1$ terms of order μ, besides terms of lower order.

Multiplying both terms of the fraction by $c - d\sqrt{Q}$, $\sqrt{Q}$ disappears from the denominator, and we may write

$$x = \frac{(ac - bdQ) + (bc - ad)\sqrt{Q}}{c^2 - d^2Q} = \alpha + \beta\sqrt{Q},$$

where α and β contain no more than $n - 1$ terms of order μ.

For a second term of order μ we have, in a similar manner, $x = \alpha_1 + \beta_1\sqrt{Q_1}$, etc.

The x *may, therefore, be transformed so as to contain a term of given order* μ *only in its numerator and there only linearly.*

We observe, however, that products of terms of order μ may occur, for α and β still depend upon $n-1$ terms of order μ. We may, then, put

$$\alpha = \alpha_{11} + \alpha_{12}\sqrt{Q_1}, \qquad \beta = \beta_{11} + \beta_{12}\sqrt{Q_1},$$

and hence

$$x = (\alpha_{11} + \alpha_{12}\sqrt{Q_1}) + (\beta_{11} + \beta_{12}\sqrt{Q_1})\sqrt{Q}.$$

6. We proceed in a similar way with the different terms of order $\mu - 1$, which occur explicitly and in Q, Q_1, etc., so that each of these quantities becomes an integral linear function of the term of order $\mu - 1$ under consideration. We then pass on to terms of lower order and finally obtain x, or rather its terms of different orders, under the form of rational integral linear functions of the individual radical expressions which occur explicitly. We then say that x is reduced to the *normal form.*

7. Let m be the total number of independent (**4**) square roots occurring in this normal form. Giving the double sign to these square roots and combining them in all possible ways, we obtain a system of 2^m values

$$x_1, x_2, \ldots . x_{2^m},$$

which we shall call *conjugate* values.

We must now investigate the equation admitting these conjugate values as roots.

8. These values are not necessarily all distinct; thus, if we have

$$x = \sqrt{a + \sqrt{b}} + \sqrt{a - \sqrt{b}},$$

this expression is not changed when we change the sign of $\sqrt{b}$.

9. If x is an arbitrary quantity and we form the polynomial

$$F(x) = (x - x_1)(x - x_2) \ldots (x - x_{2^m}),$$

$F(x) = 0$ is clearly an equation having as roots these conjugate values. It is of degree 2^m, but may have equal roots (**8**).

The coefficients of the polynomial $F(x)$ *arranged with respect to* x *are rational.*

For let us change the sign of one of the square roots; this will permute two roots, say x_λ and $x_{\lambda'}$, since the roots of $F(x) = 0$ are precisely all the conjugate values. As these roots enter $F(x)$ only under the form of the product

$$(x - x_\lambda)(x - x_{\lambda'}),$$

we merely change the order of the factors of $F(x)$. Hence the polynomial is not changed.

$F(x)$ remains, then, invariable when we change the sign of any one of the square roots; it therefore contains only their squares; and hence $F(x)$ has only rational coefficients.

10. *When any one of the conjugate values satisfies a given equation with rational coefficients,* $f(x) = 0$, *the same is true of all the others.*

$f(x)$ is not necessarily equal to $F(x)$, and may admit other roots besides the x_i's.

Let $x_1 = \alpha + \beta\sqrt{Q}$ be one of the conjugate values; $\sqrt{Q}$, a term of order μ; α and β now depend only upon other terms of order μ and terms of lower order. There must, then, be a conjugate value

$$x_1' = \alpha - \beta\sqrt{Q}.$$

Let us now form the equation $f(x_1) = 0$. $f(x_1)$ may be put into the normal form with respect to $\sqrt{Q}$,

$$f(x_1) = A + B\sqrt{Q};$$

this expression can equal zero only when A and B are simultaneously zero. Otherwise we should have

$$\sqrt{Q} = -\frac{A}{B};$$

i.e., $\sqrt{Q}$ could be expressed rationally as a function of terms of order μ and of terms of lower order contained in A and B, which is contrary to the hypothesis of the independence of all the square roots (**4**).

But we evidently have

$$f(x_1') = A - B\sqrt{Q};$$

hence if $f(x_1) = 0$, so also $f(x_1') = 0$. Whence the following proposition:

If x_1 *satisfies the equation* $f(x) = 0$, *the same is true of all the conjugate values derived from* x_1 *by changing the signs of the roots of order* μ.

The proof for the other conjugate values is obtained in an analogous manner. Suppose, for example, as may be done without affecting the generality of the reasoning, that the expression x_1 depends on only two terms of order μ, $\sqrt{Q}$ and $\sqrt{Q'}$. $f(x_1)$ may be reduced to the following normal form:

$$(a) \qquad f(x_1) = p + q\sqrt{Q} + r\sqrt{Q'} + s\sqrt{Q}\cdot\sqrt{Q'} = 0.$$

If x_1 depended on more than two terms of order μ, we should only have to add to the preceding expression a greater number of terms of analogous structure.

Equation (a) is possible only when we have separately

$$(b) \qquad p = 0, \quad q = 0, \quad r = 0, \quad s = 0.$$

Otherwise $\sqrt{Q}$ and $\sqrt{Q'}$ would be connected by a rational relation, contrary to our hypothesis.

Let now $\sqrt{R}$, $\sqrt{R'}$, ... be the terms of order $\mu - 1$ on which x_1 depends; they occur in p, q, r, s; then can the quantities p, q, r, s, in which they occur, be reduced to the normal form with respect to $\sqrt{R}$ and $\sqrt{R'}$; and if, for the sake of simplicity, we take only two quantities, $\sqrt{R}$ and $\sqrt{R'}$, we have

$$(c) \qquad p = \kappa_1 + \lambda_1 \sqrt{R} + \mu_1 \sqrt{R'} + \nu_1 \sqrt{R} \,.\, \sqrt{R'} = 0,$$

and three analogous equations for q, r, s.

The hypothesis, already used several times, of the independence of the roots, furnishes the equations

$$(d) \qquad \kappa = 0, \quad \lambda = 0, \quad \mu = 0, \quad \nu = 0.$$

Hence equations (*c*) and consequently $f(x) = 0$ are satisfied when for x_1 we substitute the conjugate values deduced by changing the signs of $\sqrt{R}$ and $\sqrt{R'}$.

Therefore *the equation* $f(x) = 0$ *is also satisfied by all the conjugate values deduced from* x_1 *by changing the signs of the roots of order* $\mu - 1$.

The same reasoning is applicable to the terms of order $\mu - 2$, $\mu - 3$, . . . and our theorem is completely proved.

11. We have so far considered two equations

$$F(x) = 0 \quad \text{and} \quad f(x) = 0.$$

Both have rational coefficients and contain the x_i's as roots. $F(x)$ is of degree 2^m and may have multiple roots ; $f(x)$ may have other roots besides the x_i's. We now introduce a third equation, $\phi(x) = 0$, defined as the equation of lowest degree, with rational coefficients, admitting the root x_1 and consequently all the x_i's (**10**).

12. Properties of the Equation $\phi(x) = 0$.

I. $\phi(x) = 0$ *is an irreducible equation, i.e.,* $\phi(x)$ cannot be resolved into two rational polynomial factors. This irreducibility is due to the hypothesis that $\phi(x) = 0$ is the rational equation of *lowest* degree satisfied by the x_i's.

For if we had

$$\phi(x) = \psi(x)\, \chi(x),$$

then $\phi(x_1) = 0$ would require either $\psi(x_1) = 0$, or $\chi(x_1) = 0$, or both. But since these equations are satisfied by all the conjugate values (**10**), $\phi(x) = 0$ would not then be the equation of lowest degree satisfied by the x_i's.

II. $\phi(x) = 0$ *has no multiple roots.* Otherwise $\phi(x)$ could be decomposed into rational factors by the well-known methods of Algebra, and $\phi(x) = 0$ would not be irreducible.

III. $\phi(x) = 0$ *has no other roots than the* x_i's. Otherwise $F(x)$ and $\phi(x)$ would admit a highest common divisor, which could be determined rationally. We could then decompose $\phi(x)$ into rational factors, and $\phi(x)$ would not be irreducible.

IV. Let M be the number of x_i's which have distinct values, and let

$$x_1, x_2, \ldots x_M$$

be these quantities. We shall then have

$$\phi(x) = C(x - x_1)(x - x_2) \ldots (x - x_M).$$

For $\phi(x) = 0$ is satisfied by the quantities x_i and it has no multiple roots. The polynomial $\phi(x)$ is then determined save for a constant factor whose value has no effect upon $\phi(x) = 0$

V. $\phi(x) = 0$ *is the only irreducible equation with rational coefficients satisfied by the* x_i's. For if $f(x) = 0$ were another rational irreducible equation satisfied by x_1 and consequently by the x_i's, $f(x)$ would be divisible by $\phi(x)$ and therefore would not be irreducible.

By reason of the five properties of $\phi(x) = 0$ thus established, we may designate this equation, in short, as *the* irreducible equation satisfied by the x_i's.

13. Let us now compare $F(x)$ and $\phi(x)$. These two polynomials have the x_i's as their only roots, and $\phi(x)$ has no multiple roots. $F(x)$ is, then, divisible by $\phi(x)$; that is,

$$F(x) = F_1(x)\,\phi(x).$$

$F_1(x)$ necessarily has rational coefficients, since it is the quotient obtained by dividing $F(x)$ by $\phi(x)$. If $F_1(x)$ is not a constant it admits roots belonging to $F(x)$; and admitting one it admits all the x_i's (**10**). Hence $F_1(x)$ is also divisible by $\phi(x)$, and

$$F_1(x) = F_2(x)\,\phi(x).$$

If $F_2(x)$ is not a constant the same reasoning still holds, the degree of the quotient being lowered by each operation. Hence at the end of a limited number of divisions we reach an equation of the form

$$F_{\nu-1}(x) = C_1 \cdot \phi(x),$$

and for $F(x)$,

$$F(x) = C_1 \cdot [\phi(x)]^\nu.$$

The polynomial $F(x)$ *is then a power of the polynomial of minimum degree* $\phi(x)$, *except for a constant factor.*

14. We can now determine the degree M of $\phi(x)$. $F(x)$ is of degree 2^m; further, it is the νth power of $\phi(x)$. Hence

$$2^m = \nu \cdot M.$$

Therefore M is also a power of 2 and we obtain the following theorem:

The degree of the irreducible equation satisfied by an expression composed of square roots only is always a power of 2

15. Since, on the other hand, there is only one irreducible equation satisfied by all the x_i's (**12**, V.), we have the converse theorem:

If an irreducible equation is not of degree 2^h, *it cannot be solved by square roots.*

CHAPTER II.

The Delian Problem and the Trisection of the Angle.

1. Let us now apply the general theorem of the preceding chapter to the *Delian problem, i.e.*, to the problem of the *duplication of the cube.* The equation of the problem is manifestly

$$x^3 = 2.$$

This is irreducible, since otherwise $\sqrt[3]{2}$ would have a rational value. For an equation of the third degree which is reducible must have a rational linear factor. Further, the degree of the equation is not of the form 2^h; hence it cannot be solved by means of square roots, and the geometric construction with straight edge and compasses is impossible.

2. Next let us consider the more general equation

$$x^3 = \lambda,$$

λ designating a parameter which may be a complex quantity of the form $a + ib$. This equation furnishes us the analytical expressions for the geometrical problems of the multiplication of the cube and the trisection of an arbitrary angle. The question arises whether this equation is reducible, *i.e.*, whether one of its roots can be expressed as a rational function of λ. It should be remarked that the irreducibility of an expression always depends upon the values of the quantities supposed to be known. In the case $x^3 = 2$, we were dealing with numerical quantities, and the question was whether $\sqrt[3]{2}$ could have a rational numerical value. In the equation $x^3 = \lambda$ we ask whether a root can be represented by a rational function of λ. In the first case, the so-called *domain of rationality* comprehends the totality of rational numbers; in the second, it is made up of the rational functions of a parameter. If no limitation is placed upon this parameter we see at once that no expression of the form $\frac{\phi(\lambda)}{\psi(\lambda)}$, in which $\phi(\lambda)$ and $\psi(\lambda)$ are polynomials, can satisfy our equation. Under our hypothesis the equation is therefore irreducible, and since its degree is not of the form 2^h, it cannot be solved by square roots.

3. Let us now restrict the variability of λ. Assume

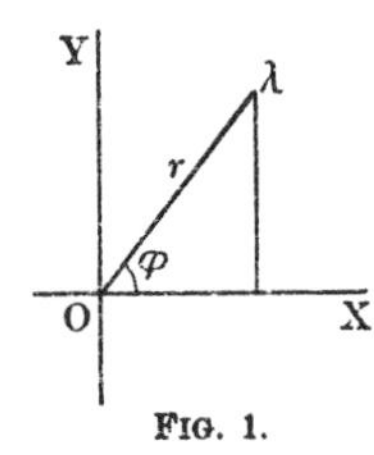

FIG. 1.

$$\lambda = r(\cos\phi + i\sin\phi);$$

whence $\sqrt[3]{\lambda} = \sqrt[3]{r}\ \sqrt[3]{\cos\phi + i\sin\phi}.$

Our problem resolves itself into two, to extract the cube root of a real number and also that of a complex number of the form $\cos\phi + i\sin\phi$, both numbers being regarded as arbitrary. We shall treat these separately.

I. The roots of the equation $x^3 = r$ are

$$\sqrt[3]{r},\ \epsilon\sqrt[3]{r},\ \epsilon^2\sqrt[3]{r},$$

representing by ϵ and ϵ^2 the complex cube roots of unity

$$\epsilon = \frac{-1 + i\sqrt{3}}{2},\quad \epsilon^2 = \frac{-1 - i\sqrt{3}}{2}.$$

Taking for the domain of rationality the totality of rational functions of r, we know by the previous reasoning that the equation $x^3 = r$ is irreducible. Hence the problem of the multiplication of the cube does not admit, in general, of a construction by means of straight edge and compasses.

II. The roots of the equation

$$x^3 = \cos\phi + i\sin\phi$$

are, by De Moivre's formula,

$$x_1 = \cos\frac{\phi}{3} + i\sin\frac{\phi}{3},$$

$$x_2 = \cos\frac{\phi+2\pi}{3} + i\sin\frac{\phi+2\pi}{3},$$

$$x_3 = \cos\frac{\phi+4\pi}{3} + i\sin\frac{\phi+4\pi}{3}.$$

These roots are represented geometrically by the vertices of an equilateral triangle inscribed in the circle with radius unity and center at the origin. The figure shows that to the root x_1 corresponds the argument $\frac{\phi}{3}$. Hence the equation

$$x^3 = \cos\phi + i\sin\phi$$

is the analytic expression of the problem of the trisection of the angle.

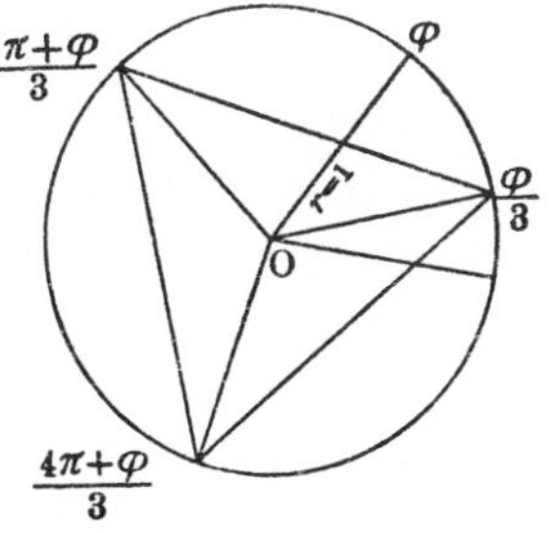

FIG. 2.

If this equation were reducible, one, at least, of its roots could be represented as a rational function of $\cos\phi$ and $\sin\phi$, its value remaining unchanged on substituting $\phi + 2\pi$ for ϕ. But if we effect this change by a continuous variation of the angle ϕ, we see that the roots x_1, x_2, x_3 undergo a cyclic permutation. *Hence no root can be represented as a rational function of* $\cos\phi$ *and* $\sin\phi$. The equation under consideration is irreducible and *therefore cannot be solved by the aid of a finite number of square roots.* Hence *the trisection of the angle cannot be effected with straight edge and compasses.*

This demonstration and the general theorem evidently hold good only when ϕ is an arbitrary angle; but for certain special values of ϕ the construction may prove to be possible, *e.g.*, when $\phi = \frac{\pi}{2}$.

CHAPTER III.

The Division of the Circle into Equal Parts.

1. The problem of dividing a given circle into n equal parts has come down from antiquity; for a long time we have known the possibility of solving it when $n = 2^h$, 3, 5, or the product of any two or three of these numbers. In his *Disquisitiones Arithmeticae*, Gauss extended this series of numbers by showing that the division is possible for every prime number of the form $p = 2^{2^\mu} + 1$ but impossible for all other prime numbers and their powers. If in $p = 2^{2^\mu} + 1$ we make $\mu = 0$ and 1, we get $p = 3$ and 5, cases already known to the ancients. For $\mu = 2$ we get $p = 2^{2^2} + 1 = 17$, a case completely discussed by Gauss.

For $\mu = 3$ we get $p = 2^{2^3} + 1 = 257$, likewise a prime number. The regular polygon of 257 sides can be constructed. Similarly for $\mu = 4$, since $2^{2^4} + 1 = 65537$ is a prime number. $\mu = 5$, $\mu = 6$, $\mu = 7$, $\mu = 8$, $\mu = 9$, $\mu = 11$, $\mu = 12$, $\mu = 15$, $\mu = 18$, $\mu = 23$, $\mu = 36$, $\mu = 38$, $\mu = 73$ give no prime numbers. The proof that the large numbers corresponding to $\mu = 5, 6, \ldots, 73$ are not prime has required a large expenditure of labor and ingenuity. It is, therefore, quite possible that $\mu = 4$ is the last number for which a solution can be effected.

Upon the regular polygon of 257 sides Richelot published an extended investigation in Crelle's *Journal*, IX, 1832, pp. 1–26, 146–161, 209–230, 337–356. The title of the memoir is: *De resolutione algebraica aequationis* $x^{257} = 1$, *sive de divisione circuli per bisectionem anguli septies repetitam in partes* 257 *inter se aequales commentatio coronata.*

To the regular polygon of 65537 sides Professor Hermes of Lingen devoted ten years of his life, examining with care all the roots furnished by Gauss's method. His MSS. are preserved in the collection of the mathematical seminary in Göttingen. (Compare a communication of Professor Hermes in No. 3 of the *Göttinger Nachrichten* for 1894.)

2. We may restrict the problem of the division of the circle into n equal parts to the cases where n is a prime number p or a power p^α of such a number. For if n is a composite number and if μ and ν are factors of n, prime to each other, we can always find integers a and b, positive or negative, such that

$$1 = a\mu + b\nu;$$

whence

$$\frac{1}{\mu\nu} = \frac{a}{\nu} + \frac{b}{\mu}.$$

To divide the circle into $\mu\nu = n$ equal parts it is sufficient to know how to divide it into μ and ν equal parts respectively. Thus, for $n = 15$, we have

$$\frac{1}{15} = \frac{2}{3} - \frac{3}{5}.$$

3. As will appear, the division into p equal parts (p being a prime number) is possible only when p is of the form $p = 2^h + 1$. We shall next show that a prime number can be of this form only when $h = 2^\mu$. For this we shall make use of Fermat's Theorem :

If p *is a prime number and* a *an integer not divisible by* p, *these numbers satisfy the congruence*

$$a^{p-1} \equiv +1 \pmod{p}.$$

$p - 1$ is not necessarily the lowest exponent which, for a given value of a, satisfies the congruence. If s is the lowest exponent it may be shown that s is a divisor of $p - 1$. In particular, if $s = p - 1$ we say that a is a *primitive root* of p, and notice that for every prime number p there is a primitive root. We shall make use of this notion further on.

Suppose, then, p a prime number such that

$$(1) \qquad p = 2^h + 1,$$

and s the least integer satisfying

$$(2) \qquad 2^s \equiv +1 \pmod{p}.$$

From (1) $2^h < p$; from (2) $2^s > p$.

$$\therefore s > h.$$

(1) shows that h is the least integer satisfying the congruence

$$(3) \qquad 2^h \equiv -1 \pmod{p}.$$

From (2) and (3), by division,

$$2^{s-h} \equiv -1 \pmod{p}.$$

$$\therefore (4) \qquad s - h \nless h, \quad s \nless 2h.$$

From (3), by squaring,

$$2^{2h} \equiv 1 \pmod{p}.$$

Comparing with (2) and observing that s is the least exponent satisfying congruences of the form

$$2^x \equiv 1 \pmod{p},$$

we have

$$(5) \qquad s \ngtr 2h.$$

$$\therefore s = 2h.$$

We have observed that s is a divisor of $p - 1 = 2^h$; the same is true of h, which is, therefore, a power of 2. Hence prime numbers of the form $2^h + 1$ are necessarily of the form $2^{2^\mu} + 1$.

4. This conclusion may be established otherwise. Suppose that h is divisible by an odd number, so that

$$h = h'(2r + 1);$$

then, by reason of the formula

$$x^{2n+1} + 1 = (x + 1)(x^{2n} - x^{2n-1} + \ldots - x + 1),$$

$p = 2^{h'(2n+1)} + 1$ is divisible by $2^{h'} + 1$, and hence is not a prime number.

5. We now reach our fundamental proposition :

p *being a prime number, the division of the circle into* p *equal parts by the straight edge and compasses is impossible unless* p *is of the form*

$$p = 2^h + 1 = 2^{2^\mu} + 1.$$

Let us trace in the z-plane $(z = x + iy)$ a circle of radius 1. To divide this circle into n equal parts, beginning at $z = 1$, is the same as to solve the equation

$$z^n - 1 = 0.$$

This equation admits the root $z = 1$; let us suppress this root by dividing by $z - 1$, which is the same geometrically as to disregard the initial point of the division. We thus obtain the equation

$$z^{n-1} + z^{n-2} + \ldots + z + 1 = 0,$$

which may be called the *cyclotomic equation.* As noticed above, we may confine our attention to the cases where n is a prime number or a power of a prime number. We shall first investigate the case when $n = p$. The essential point of the proof is to show *that the above equation is irreducible.* For since, as we have seen, irreducible equations can only be solved by means of square roots in finite number when their degree is a power of 2, a division into p parts is always impossible when $p - 1$ is not equal to a power of 2, *i.e.*. when

$$p \neq 2^h + 1 \neq 2^{2^\mu} + 1.$$

Thus we see why Gauss's prime numbers occupy such an exceptional position.

6. At this point we introduce a lemma known as *Gauss's Lemma.* If

$$F(z) = z^m + Az^{m-1} + Bz^{m-2} + \ldots + Lz + M,$$

where A, B, . . . are integers, and $F(z)$ can be resolved into two rational factors $f(z)$ and $\phi(z)$, so that

$$F(z) = f(z) \cdot \phi(z) = (z^{m'} + \alpha_1 z^{m'-1} + \alpha_2 z^{m'-2} + \ldots)$$
$$(z^{m''} + \beta_1 z^{m''-1} + \beta_2 z^{m''-2} + \ldots),$$

then must the α's and β's also be integers. In other words :

If an integral expression can be resolved into rational factors these factors must be integral expressions.

Let us suppose the α's and β's to be fractional. In each factor reduce all the coefficients to the least common denominator. Let a_0 and b_0 be these common denominators. Finally multiply both members of our equation by $a_0 b_0$. It takes the form

$$a_0 b_0 F(z) = f_1(z)\, \phi_1(z) = (a_0 z^{m'} + a_1 z^{m'-1} + \ldots)$$
$$(b_0 z^{m''} + b_1 z^{m''-1} + \ldots).$$

The a's are integral and prime to one another, as also the b's, since a_0 and b_0 are the least common denominators.

Suppose a_0 and b_0 different from unity and let q be a prime divisor of $a_0 b_0$. Further, let a_i be the first coefficient of $f_1(z)$ and b_k the first coefficient of $\phi_1(z)$ not divisible by q. Let us develop the product $f_1(z)\, \phi_1(z)$ and consider the coefficient of $z^{m'+m''-i-k}$. It will be

$$a_i b_k + a_{i-1} b_{k+1} + a_{i-2} b_{k+2} + \ldots + a_{i+1} b_{k-1} + a_{i+2} b_{k-2} + \ldots$$

According to our hypotheses, all the terms after the first are divisible by q, but the first is not. Hence this coefficient is not divisible by q. Now the coefficient of $z^{m'+m''-i-k}$ in the first member is divisible by $a_0 b_0$, *i.e.*, by q. Hence if the identity is true it is impossible for a coefficient not divisible by q to occur in each polynomial. The coefficients of one at least of the polynomials are then all divisible by q. Here is another absurdity, since we have seen that all the coefficients are prime to one another. Hence we cannot suppose a_0 and b_0 different from 1, and consequently the α's and β's are integral.

7. In order to show that the cyclotomic equation is irreducible, it is sufficient to show by Gauss's Lemma that the first member cannot be resolved into factors with integral coefficients. To this end we shall employ the simple method due to Eisenstein, in Crelle's *Journal*, XXXIX, p. 167, which depends upon the substitution

$$z = x + 1.$$

We obtain

$$f(z) = \frac{z^p - 1}{z - 1} = \frac{(x+1)^p - 1}{x} = x^{p-1} + px^{p-2} + \frac{p(p-1)}{1 \cdot 2} x^{p-3}$$
$$+ \ldots + \frac{p(p-1)}{1 \cdot 2} x + p = 0.$$

All the coefficients of the expanded member except the first are divisible by p; the last coefficient is always p itself, by hypothesis a prime number. An expression of this class is always irreducible.

For if this were not the case we should have

$$f(x+1) = (x^m + a_1 x^{m-1} + \ldots + a_{m-1} x + a_m)$$
$$(x^{m'} + b_1 x^{m'-1} + \ldots + b_{m'-1} x + b_{m'}),$$

where the a's and b's are integers.

Since the term of zero degree in the above expression of $f(z)$ is p, we have $a_m b_{m'} = p$. p being prime, one of the factors of $a_m b_{m'}$ must be unity. Suppose, then,

$$a_m = \pm p, \; b_m = \pm 1.$$

Equating the coefficients of the terms in x, we have

$$\frac{p(p-1)}{2} = a_{m-1} b_{m'} + a_m b_{m'-1}.$$

The first member and the second term of the second being divisible by p, $a_{m-1} b_m$ must be so also. Since $b_m = \pm 1$, a_{m-1} is divisible by p. Equating the coefficients of the terms in x^2 we may show that a_{m-2} is divisible by p. Similarly we show that all of the remaining coefficients of the factor $x^m + a_1 x^{m-1} + \ldots + a_{m-1} x + a_m$ are divisible by p. But this cannot be true of the coefficient of x^m, which is 1. The assumed equality is impossible and hence the cyclotomic equation is irreducible when p is a prime.

8. We now consider the case where n is a power of a prime number, say $n = p^\alpha$. We propose to show that when $p > 2$ the division of the circle into p^2 equal parts is impossible. The general problem will then be solved, since the division into p^α equal parts evidently includes the division into p^2 equal parts.

The cyclotomic equation is now

$$\frac{z^{p^2} - 1}{z - 1} = 0.$$

It admits as roots extraneous to the problem those which come from the division into p equal parts, *i.e.*, the roots of the equation

$$\frac{z^p - 1}{z - 1} = 0.$$

Suppressing these roots by division we obtain

$$f(z) = \frac{z^{p^2} - 1}{z^p - 1} = 0$$

as the cyclotomic equation. This may be written

$$z^{p(p-1)} + z^{p(p-2)} + \ldots + z^p + 1 = 0.$$

Transforming by the substitution

$$z = x + 1,$$

we have

$$(x + 1)^{p(p-1)} + (x + 1)^{p(p-2)} + \ldots + (x + 1)^p + 1 = 0.$$

The number of terms being p, the term independent of x after development will be equal to p, and the sum will take the form

$$x^{p(p-1)} + p \cdot \chi(x),$$

where $\chi(x)$ is a polynomial with integral coefficients whose constant term is 1. We have just shown that such an expression is always irreducible. Consequently *the new cyclotomic equation is also irreducible.*

The degree of this equation is $p(p-1)$. On the other hand an irreducible equation is solvable by square roots only when its degree is a power of 2. Hence a circle is divisible into p^2 equal parts only when $p = 2$, p being assumed to be a prime.

The same is true, as already noted, for the division into p^α equal parts when $\alpha > 2$.

CHAPTER IV.

The Construction of the Regular Polygon of 17 Sides.

1. We have just seen that the division of the circle into equal parts by the straight edge and compasses is possible only for the prime numbers studied by Gauss. It will now be of interest to learn how the construction can actually be effected.

The purpose of this chapter, then, will be to show in an elementary way how to inscribe in the circle the regular polygon of 17 sides.

Since we possess as yet no method of construction based upon considerations purely geometrical, we must follow the path indicated by our general discussions. We consider, first of all, the roots of the cyclotomic equation

$$x^{16} + x^{15} + \ldots + x^2 + x + 1 = 0,$$

and construct geometrically the expression, formed of square roots, deduced from it.

We know that the roots can be put into the transcendental form

$$\epsilon_\kappa = \cos\frac{2\kappa\pi}{17} + i \sin\frac{2\kappa\pi}{17} \quad (\kappa = 1, 2, \ldots 16);$$

and if

$$\epsilon_1 = \cos\frac{2\pi}{17} + i \sin\frac{2\pi}{17},$$

that

$$\epsilon_\kappa = \epsilon_1{}^\kappa.$$

Geometrically, these roots are represented in the complex plane by the vertices, different from 1, of the regular polygon of 17 sides inscribed in a circle of radius 1, having the origin as center. The selection of ϵ_1 is arbitrary, but for the construction it is essential to indicate some ϵ as the point of departure. Having fixed upon ϵ_1, the angle corresponding to ϵ_κ is κ times the angle corresponding to ϵ_1, which completely determines ϵ_κ.

2. The fundamental idea of the solution is the following: *Forming a primitive root to the modulus* 17 *we may arrange the* 16 *roots of the equation in a cycle in a determinate order.*

As already stated, a number a is said to be a primitive root to the modulus 17 when the congruence

$$a^s \equiv +1 \pmod{17}$$

has for least solution $s = 17 - 1 = 16$. The number 3 possesses this property; for we have

$$\left.\begin{array}{llll} 3^1 \equiv 3 & 3^5 \equiv 5 & 3^9 \equiv 14 & 3^{13} \equiv 12 \\ 3^2 \equiv 9 & 3^6 \equiv 15 & 3^{10} \equiv 8 & 3^{14} \equiv 2 \\ 3^3 \equiv 10 & 3^7 \equiv 11 & 3^{11} \equiv 7 & 3^{15} \equiv 6 \\ 3^4 \equiv 13 & 3^8 \equiv 16 & 3^{12} \equiv 4 & 3^{16} \equiv 1 \end{array}\right\} \pmod{17}.$$

Let us then arrange the roots ϵ_κ so that their subscripts are the preceding remainders in order

$$\epsilon_3, \epsilon_9, \epsilon_{10}, \epsilon_{13}, \epsilon_5, \epsilon_{15}, \epsilon_{11}, \epsilon_{16}, \epsilon_{14}, \epsilon_8, \epsilon_7, \epsilon_4, \epsilon_{12}, \epsilon_2, \epsilon_6, \epsilon_1.$$

Notice that if r is the remainder of 3^κ (mod. 17), we have

$$3^\kappa = 17q + r,$$

whence
$$\epsilon_r = \epsilon_1{}^r = \epsilon_1{}^{3^\kappa}$$

If r′ is the next remainder, we have similarly

$$\epsilon_{r'} = \epsilon_1{}^{3^{\kappa+1}} = (\epsilon_1{}^{3^\kappa})^3 = (\epsilon_r)^3.$$

Hence in this series of roots each root is the cube of the preceding.

Gauss's method consists in decomposing this cycle into sums containing 8, 4, 2, 1 roots respectively, corresponding to the divisors of 16. Each of these sums is called a period.

The periods thus obtained may be calculated successively as roots of certain quadratic equations.

The process just outlined is only a particular case of that employed in the general case of the division into p equal parts. The $p-1$ roots of the cyclotomic equation are cyclically arranged by means of a primitive root of p, and the periods may be calculated as roots of certain auxiliary equations. The degree of these last depends upon the prime factors of $p-1$. They are not necessarily equations of the second degree.

The general case has, of course, been treated in detail by Gauss in his *Disquisitiones*, and also by Bachmann in his work, *Die Lehre von der Kreisteilung* (Leipzig, 1872).

8. In our case of the 16 roots the periods may be formed in the following manner: Form two periods of 8 roots by taking in the cycle, first, the roots of even order, then those of odd order. Designate these periods by x_1 and x_2, and replace each root by its index. We may then write symbolically

$$x_1 = 9 + 13 + 15 + 16 + 8 + 4 + 2 + 1,$$
$$x_2 = 3 + 10 + 5 + 11 + 14 + 7 + 12 + 6.$$

Operating upon x_1 and x_2 in the same way, we form 4 periods of 4 terms:

$$y_1 = 13 + 16 + 4 + 1,$$
$$y_2 = 9 + 15 + 8 + 2,$$
$$y_3 = 10 + 11 + 7 + 6,$$
$$y_4 = 3 + 5 + 14 + 12.$$

Operating in the same way upon the y's, we obtain 8 periods of 2 terms:

$$z_1 = 16 + 1, \qquad z_5 = 11 + 6,$$
$$z_2 = 13 + 4, \qquad z_6 = 10 + 7,$$
$$z_3 = 15 + 2, \qquad z_7 = 5 + 12,$$
$$z_4 = 9 + 8, \qquad z_8 = 3 + 14.$$

It now remains to show that *these periods can be calculated successively by the aid of square roots.*

4. It is readily seen that the sum of the remainders corresponding to the roots forming a period z is always equal to 17. These roots are then ϵ_r and ϵ_{17-r};

$$\epsilon_r = \cos r\frac{2\pi}{17} + i \sin r\frac{2\pi}{17},$$

$$\epsilon_1 = \epsilon_{17-r} = \cos (17 - r)\frac{2\pi}{17} + i \sin (17 - r)\frac{2\pi}{17},$$

$$= \cos r\frac{2\pi}{17} - i \sin r\frac{2\pi}{17}.$$

Hence

$$\epsilon_r + \epsilon_{r'} = 2 \cos r\frac{2\pi}{17}.$$

Therefore all the periods z are real, and we readily obtain

$$z_1 = 2 \cos \frac{2\pi}{17}, \qquad z_5 = 2 \cos 6\frac{2\pi}{17},$$

$$z_2 = 2 \cos 4\frac{2\pi}{17}, \qquad z_6 = 2 \cos 7\frac{2\pi}{17},$$

$$z_3 = 2 \cos 2\frac{2\pi}{17}, \qquad z_7 = 2 \cos 5\frac{2\pi}{17},$$

$$z_4 = 2 \cos 8\frac{2\pi}{17}, \qquad z_8 = 2 \cos 3\frac{2\pi}{17}.$$

Moreover, by definition,

$$x_1 = z_1 + z_2 + z_3 + z_4, \qquad x_2 = z_5 + z_6 + z_7 + z_8,$$

$$y_1 = z_1 + z_2, \quad y_2 = z_3 + z_4, \quad y_3 = z_5 + z_6, \quad y_4 = z_7 + z_8.$$

5. It will be necessary to determine the relative magnitude of the different periods. For this purpose we shall employ the following artifice: We divide the semicircle of unit radius into 17 equal parts and denote by $S_1, S_2, \ldots S_{17}$ the distances of the consecutive points of division $A_1, A_2, \ldots A_{17}$ from the initial point of the semicircle, S_{17} being equal to the diameter, *i.e.*, equal to 2. The angle $A_\kappa A_{17} O$ has the same measure as the half of the arc $A_\kappa O$, which equals $\frac{2\kappa\pi}{34}$. Hence

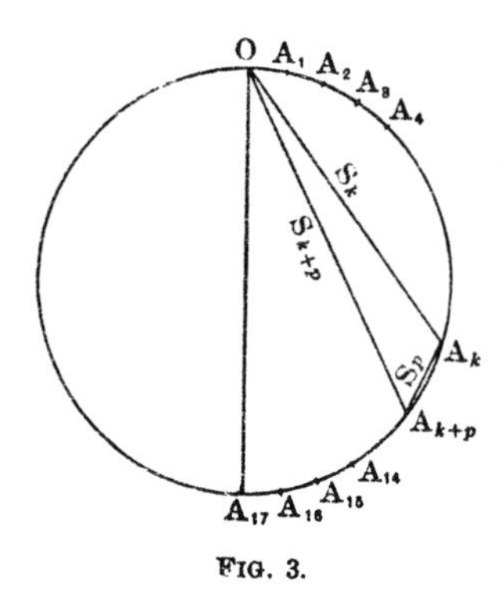

FIG. 3.

$$S_\kappa = 2 \sin \frac{\kappa\pi}{34} = 2 \cos \frac{(17-\kappa)\pi}{34}.$$

That this may be identical with $2 \cos h \frac{2\pi}{17}$, we must have

$$4h = 17 - \kappa,$$
$$\kappa = 17 - 4h.$$

Giving to h the values 1, 2, 3, 4, 5, 6, 7, 8, we find for κ the values 13, 9, 5, 1, -3, -7, -11, -15. Hence

$$z_1 = S_{13}, \qquad z_5 = -S_7,$$
$$z_2 = S_1, \qquad z_6 = -S_{11},$$
$$z_3 = S_9, \qquad z_7 = -S_3,$$
$$z_4 = -S_{15}, \qquad z_8 = S_5.$$

The figure shows that S_κ increases with the subscript; hence the order of increasing magnitude of the periods z is

$$z_4, z_6, z_5, z_7, z_2, z_8, z_3, z_1.$$

Moreover, the chord $A_\kappa A_{\kappa+p}$ subtends p divisions of the semicircumference and is equal to S_p; the triangle $OA_\kappa A_{\kappa+p}$ shows that

$$S_{\kappa+p} < S_\kappa + S_p,$$

and *a fortiori*

$$S_{\kappa+p} < S_{\kappa+r} + S_{p+r}.$$

Calculating the differences two and two of the periods y, we easily find

$$y_1 - y_2 = S_{13} + S_1 - S_9 + S_{15} > 0,$$
$$y_1 - y_3 = S_{13} + S_1 + S_7 + S_{11} > 0,$$
$$y_1 - y_4 = S_{13} + S_1 + S_3 - S_5 > 0,$$
$$y_2 - y_3 = S_9 - S_{15} + S_7 + S_{11} > 0,$$
$$y_2 - y_4 = S_9 - S_{15} + S_3 - S_5 < 0,$$
$$y_3 - y_4 = - S_7 - S_{11} + S_3 - S_5 < 0.$$

Hence

$$y_3 < y_2 < y_4 < y_1.$$

Finally we obtain in a similar way

$$x_2 < x_1.$$

6. We now propose to calculate $z_1 = 2 \cos \frac{2\pi}{17}$. After making this calculation and constructing z_1, we can easily deduce the side of the regular polygon of 17 sides. In order to find the quadratic equation satisfied by the periods, we proceed to determine symmetric functions of the periods.

Associating z_1 with the period z_2 and thus forming the period y_1, we have, first,

$$z_1 + z_2 = y_1.$$

Let us now determine $z_1 z_2$. We have

$$z_1 z_2 = (16 + 1)(13 + 4),$$

where the symbolic product κp represents

$$\epsilon_\kappa \cdot \epsilon_p = \epsilon_{\kappa + p}.$$

Hence it should be represented symbolically by $\kappa + p$, remembering to subtract 17 from $\kappa + p$ as often as possible. Thus,

$$z_1 z_2 = 12 + 3 + 14 + 5 = y_4.$$

Therefore z_1 and z_2 are the roots of the quadratic equation

$$(\zeta) \qquad z^2 - y_1 z + y_4 = 0,$$

whence, since $z_1 > z_2$,

$$z_1 = \frac{y_1 + \sqrt{y_1^2 - 4y_4}}{2}, \qquad z_2 = \frac{y_1 - \sqrt{y_1^2 - 4y_4}}{2}.$$

We must now determine y_1 and y_4. Associating y_1 with the period y_2, thus forming the period x_1, and y_3 with the period y_4, thus forming the period x_2, we have, first,

$$y_1 + y_2 = x_1.$$

Then,

$$y_1 y_2 = (13 + 16 + 4 + 1)(9 + 15 + 8 + 2).$$

Expanding symbolically, the second member becomes equal to the sum of all the roots; that is, to -1. Therefore y_1 and y_2 are the roots of the equation

$$(\eta) \qquad y^2 - x_1 y - 1 = 0,$$

whence, since $y_1 > y_2$,

$$y_1 = \frac{x_1 + \sqrt{x_1^2 + 4}}{2}, \qquad y_2 = \frac{x_1 - \sqrt{x_1^2 + 4}}{2}$$

Similarly,

$$y_3 + y_4 = x_2$$

and

$$y_3 y_4 = -1.$$

Hence y_3 and y_4 are the roots of the equation

$$(\eta') \qquad y^2 - x_2 y - 1 = 0;$$

whence, since $y_4 > y_3$,

$$y_4 = \frac{x_2 + \sqrt{x_2^2 + 4}}{2}, \qquad y_3 = \frac{x_2 - \sqrt{x_2^2 + 4}}{2}$$

It now remains to determine x_1 and x_2. Since $x_1 + x_2$ is equal to the sum of all the roots,

$$x_1 + x_2 = -1.$$

Further,

$$x_1 x_2 = (13 + 16 + 4 + 1 + 9 + 15 + 8 + 2)$$
$$(10 + 11 + 7 + 6 + 3 + 5 + 14 + 12).$$

Expanding symbolically, each root occurs 4 times, and thus

$$x_1 x_2 = -4.$$

Therefore x_1 and x_2 are the roots of the quadratic

$$(\xi) \qquad x^2 + x - 4 = 0;$$

whence, since $x_1 > x_2$,

$$x_1 = \frac{-1 + \sqrt{17}}{2}, \qquad x_2 = \frac{-1 - \sqrt{17}}{2}$$

Solving equations ξ, η, η', ζ in succession, z_1 is determined by a series of square roots.

Effecting the calculations, we see that z_1 depends upon the four square roots

$$\sqrt{17}, \ \sqrt{x_1^2 + 4}, \ \sqrt{x_2^2 + 4}, \ \sqrt{y_1^2 - 4y_4}.$$

If we wish to reduce z_1 to the normal form we must see whether any one of these square roots can be expressed rationally in terms of the others.

Now, from the roots of (η),

$$\sqrt{x_1^2 + 4} = y_1 - y_2,$$

$$\sqrt{x_2^2 + 4} = y_4 - y_3.$$

Expanding symbolically, we verify that

$$(y_1 - y_2)(y_4 - y_3) = 2(x_1 - x_2),^*$$

* $(y_1 - y_2)(y_4 - y_3) = (13 + 16 + 4 + 1 - 9 - 15 - 8 - 2)(3 + 5 + 14 + 12 - 10 - 11 - 7 - 6)$

$$\begin{aligned} &= 16 + 1 + 10 + 8 - 6 - 7 - 3 - 2 \\ &+ 2 + 4 + 13 + 11 - 9 - 10 - 6 - 5 \\ &+ 7 + 9 + 1 + 16 - 14 - 15 - 11 - 10 \\ &+ 4 + 6 + 15 + 13 - 11 - 12 - 8 - 7 \\ &- 12 - 14 - 6 - 4 + 2 + 3 + 16 + 15 \\ &- 1 - 3 - 12 - 10 + 8 + 9 + 5 + 4 \\ &- 11 - 13 - 5 - 3 + 1 + 2 + 15 + 14 \\ &- 5 - 7 - 16 - 14 + 12 + 13 + 9 + 8 \end{aligned}$$

$= 2(16 + 1 + 8 + 2 + 4 + 13 + 15 + 9 - 10 - 6 - 7 - 3 - 11 - 5 - 14 - 12)$

$= 2(x_1 - x_2)$.

that is,

$$\sqrt{x_1^2+4}\ \sqrt{x_2^2+4}=2\sqrt{17}.$$

Hence $\sqrt{x_2^2+4}$ can be expressed rationally in terms of the other two square roots. This equation shows that if two of the three differences y_1-y_2, y_4-y_3, x_1-x_2 are positive, the same is true of the third, which agrees with the results obtained directly.

Replacing now x_1, y_1, y_4 by their numerical values, we obtain in succession

$$x_1=\frac{-1+\sqrt{17}}{2},$$

$$y_1=\frac{-1+\sqrt{17}+\sqrt{34-2\sqrt{17}}}{4},$$

$$y_4=\frac{-1-\sqrt{17}+\sqrt{34+2\sqrt{17}}}{4},$$

$$z_1=\frac{-1+\sqrt{17}+\sqrt{34-2\sqrt{17}}}{8}$$

$$+\frac{\sqrt{68+12\sqrt{17}-16\sqrt{34+2\sqrt{17}}-2(1-\sqrt{17})\sqrt{34-2\sqrt{17}}}}{8}$$

The algebraic part of the solution of our problem is now completed. We have already remarked that there is no known construction of the regular polygon of 17 sides based upon purely geometric considerations. There remains, then, only the geometric translation of the individual algebraic steps.

7. We may be allowed to introduce here a brief historical account of geometric constructions with straight edge and compasses.

In the geometry of the ancients the straight edge and compasses were always used together; the difficulty lay merely in bringing together the different parts of the figure so as not to draw any unnecessary lines. Whether the several steps in the construction were made with straight edge or with compasses was a matter of indifference.

On the contrary, in 1797, the Italian Mascheroni succeeded in effecting all these constructions with the compasses alone; he set forth his methods in his *Geometria del compasso*, and claimed that constructions with compasses were practically more exact than those with the straight edge. As he expressly stated, he wrote for mechanics, and therefore with a practical end in view. Mascheroni's original work is difficult to read, and we are under obligations to Hutt for furnishing a brief *résumé* in German, *Die Mascheroni'schen Constructionen* (Halle, 1880).

Soon after, the French, especially the disciples of Carnot, the author of the *Géométrie de position*, strove, on the other hand, to effect their constructions as far as possible with the straight edge. (See also Lambert, *Freie Perspective*, 1774.)

Here we may ask a question which algébra enables us to answer immediately: In what cases can the solution of an algebraic problem be constructed with the straight edge alone? The answer is not given with sufficient explicitness by the authors mentioned. We shall say:

With the straight edge alone we can construct all algebraic expressions whose form is rational.

With a similar view Brianchon published in 1818 a paper, *Les applications de la théorie des transversales,* in which he shows how his constructions can be effected in many cases with the straight edge alone. He likewise insists upon the practical value of his methods, which are especially adapted to field work in surveying.

Poncelet was the first, in his *Traité des propriétés projectives* (Vol. I, Nos. 351–357), to conceive the idea that it is sufficient to use a *single fixed circle* in connection with the straight lines of the plane in order to construct all expressions depending upon square roots, the center of the fixed circle being given.

This thought was developed by Steiner in 1833 in a celebrated paper entitled *Die geometrischen Constructionen, ausgeführt mittels der geraden Linie und eines festen Kreises, als Lehrgegenstand für höhere Unterrichtsanstalten und zum Selbstunterricht.*

8. To construct the regular polygon of 17 sides we shall follow the method indicated by von Staudt (Crelle's *Journal*, Vol. XXIV, 1842), modified later by Schröter (Crelle's *Journal*, Vol. LXXV, 1872). The construction of the regular polygon of 17 sides is made in accordance with the methods indicated by Poncelet and Steiner, inasmuch as besides the straight edge but one fixed circle is used.*

First, we will show *how with the straight edge and one fixed circle we can solve every quadratic equation.*

At the extremities of a diameter of the fixed unit circle (Fig. 4) we draw two tangents, and select the lower as the axis of X, and the diameter perpendicular to it as the axis of Y. Then the equation of the circle is

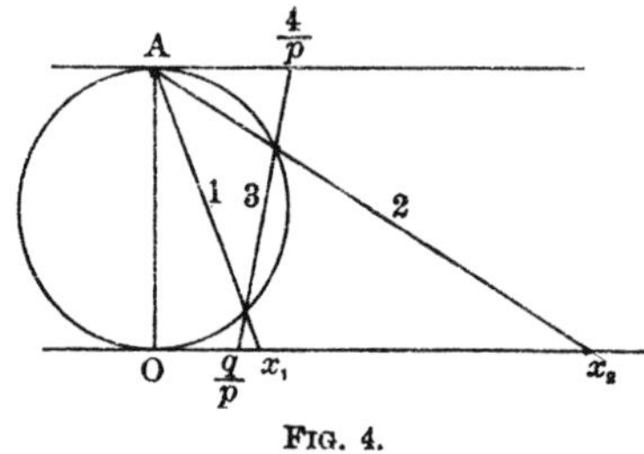

FIG. 4.

$$x^2 + y(y-2) = 0.$$

Let

$$x^2 - px + q = 0$$

be any quadratic equation with real roots x_1 and x_2. Required to construct the roots x_1 and x_2 upon the axis of X.

Lay off upon the upper tangent from A to the right, a segment measured by $\frac{4}{p}$; upon the axis of X from O, a segment

* A Mascheroni construction of the regular polygon of 17 sides by L. Gérard is given in *Math. Annalen*, Vol. XLVIII, 1896, pp. 390–392.

measured by $\frac{q}{p}$; connect the extremities of these segments by the line 3 and project the intersections of this line with the circle from A, by the lines 1 and 2, upon the axis of X. The segments thus cut off upon the axis of X are measured by x_1 and x_2.

Proof. Calling the intercepts upon the axis of X, x_1 and x_2, we have the equation of the line 1,

$$2x + x_1(y-2) = 0;$$

of the line 2.

$$2x + x_2(y-2) = 0.$$

If we multiply the first members of these two equations we get

$$x^2 + \frac{x_1 + x_2}{2} x(y-2) + \frac{x_1 x_2}{4}(y-2)^2 = 0$$

as the equation of the line pair formed by 1 and 2. Subtracting from this the equation of the circle, we obtain

$$\frac{x_1 + x_2}{2} x(y-2) + \frac{x_1 x_2}{4}(y-2)^2 - y(y-2) = 0$$

This is the equation of a conic passing through the four intersections of the lines 1 and 2 with the circle. From this equation we can remove the factor $y-2$, corresponding to the tangent, and we have left

$$\frac{x_1 + x_2}{2} x + \frac{x_1 x_2}{4}(y-2) - y = 0,$$

which is the equation of the line 3. If we now make $x_1 + x_2 = p$ and $x_1 x_2 = q$, we get

$$\frac{p}{2} x + \frac{q}{4}(y-2) - y = 0,$$

and the transversal 3 cuts off from the line $y = 2$ the segment $\frac{4}{p}$, and from the line $y=0$ the segment $\frac{q}{p}$. Thus the correctness of the construction is established.

9. In accordance with the method just explained, we shall now construct the roots of our four quadratic equations. They are (see pp. 29–31)

(ξ) $x^2 + x - 4 = 0$, with roots x_1 and x_2; $x_1 > x_2$,
(η) $y^2 - x_1 y - 1 = 0$, with roots y_1 and y_2; $y_1 > y_2$,
(η') $y^2 - x_2 y - 1 = 0$, with roots y_3 and y_4; $y_4 > y_3$,
(ζ) $z^2 - y_1 z + y_4 = 0$, with roots z_1 and z_2; $z_1 > z_2$.
These will furnish

$$z_1 = 2 \cos \frac{2\pi}{17},$$

whence it is easy to construct the polygon desired. We notice further that to construct z_1 it is sufficient to construct x_1, x_2, y_1, y_4.

We then lay off the following segments: upon the upper tangent, $y=2$,

$$-4, \frac{4}{x_1}, \frac{4}{x_2}, \frac{4}{y_1};$$

upon the axis of X,

$$+4, -\frac{1}{x_1}, -\frac{1}{x_2}, \frac{y_4}{y_1}.$$

This may all be done in the following manner: The straight line connecting the point $+4$ upon the axis of X with the point -4 upon the tangent $y=2$ cuts the circle in

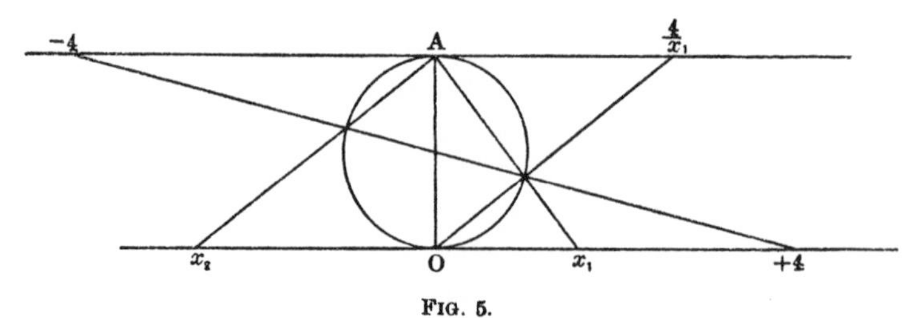

FIG. 5.

two points, the projection of which from the point A (0, 2), the upper vertex of the circle, gives the two roots x_1, x_2 of the first quadratic equation as intercepts upon the axis of X.

To solve the second equation we have to lay off $\frac{4}{x_1}$ above and $-\frac{1}{x_1}$ below.

To determine the first point we connect x_1 upon the axis of X with A, the upper vertex, and from O, the lower vertex, draw another straight line through the intersection of this line with the circle. This cuts off upon the upper tangent the intercept $\frac{4}{x_1}$. This can easily be shown analytically.

The equation of the line from A to x_1 (Fig. 5),

$$2x + x_1 y = 2x_1,$$

and that of the circle,

$$x^2 + y(y - 2) = 0,$$

give as the coördinates of their intersection

$$\frac{4x_1}{x_1^2 + 4}, \quad \frac{2x_1^2}{x_1^2 + 4}.$$

The equation of the line from O through this point becomes

$$y = \frac{x_1}{2} x,$$

cutting off upon $y = 2$ the intercept $\frac{4}{x_1}$.

We reach the same conclusion still more simply by the use of some elementary notions of projective geometry. By our construction we have obviously associated with every point x of the lower range one, and only one, point of the upper, so that to the point $x = \infty$ corresponds the point $x' = 0$, and conversely. Since in such a correspondence there must exist a linear relation, the abscissa x' of the upper point must satisfy the equation

$$x' = \frac{\text{const.}}{x}$$

Since $x' = 2$ when $x = 2$, as is obvious from the figure, the constant $= 4$.

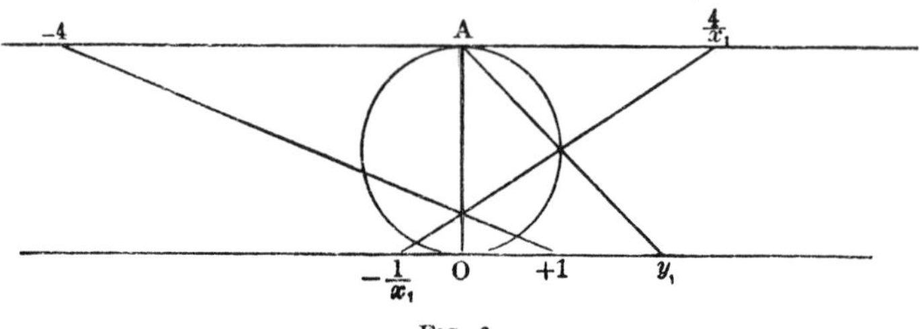

FIG. 6.

To determine $-\frac{1}{x_1}$ upon the axis of X we connect the point -4 upon the upper with the point $+1$ upon the lower tangent (Fig. 6). The point thus determined upon the vertical diameter we connect with the point $\frac{4}{x_1}$ above. This line cuts off upon the axis of X the intercept $-\frac{1}{x_1}$. For the line from -4 to $+1$,

$$5y + 2x = 2,$$

intersects the vertical diameter in the point $(0, \frac{2}{5})$. Hence the equation of the line from $\frac{4}{x_1}$ to this point is

$$5y - 2x_1x = 2,$$

and its intersection with the lower tangent gives $-\frac{1}{x_1}$

The projection from A of the intersections of the line from $-\frac{1}{x_1}$ to $\frac{4}{x_1}$ with the circle determines upon the axis of X the two roots of the second quadratic equation, of which, as already noted, we need only the greater, y_1. This corresponds, as shown by the figure, to the projection of the upper intersection of our transversal with the circle.

Similarly, we obtain the roots of the third quadratic equation. Upon the upper tangent we project from O the intersection of the circle with the straight line which gave upon the axis of X the root $+x_2$. This immediately gives the intercept $\frac{4}{x_2}$, by reason of the correspondence just explained.

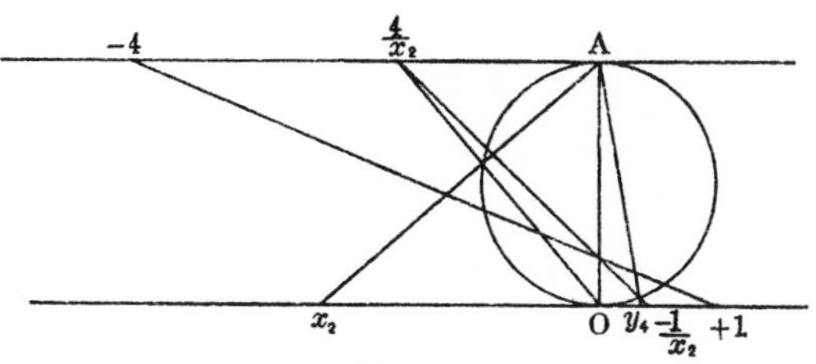

FIG. 7.

If we connect this point with the point where the vertical diameter intersects the line joining -4 above and $+1$ below, we cut off upon the axis of X the segment $-\frac{1}{x_2}$, as desired. If we project that intersection of this transversal with the circle which lies in the positive quadrant from A upon the axis of X, we have constructed the required root y_4 of the third quadratic equation.

We have finally to determine the root z_1 of the fourth quadratic equation and for this purpose to lay off $\frac{4}{y_1}$ above and $\frac{y_4}{y_1}$ below. We solve the first problem in the usual way, by projecting the intersection of the circle with the line connecting A with $+y_1$ below, from O upon the upper tangent, thus obtaining $\frac{4}{y_1}$. For the other segment we connect the point $+4$ above with y_4 below, and then the point thus determined upon the vertical diameter produced with $\frac{4}{y_1}$. This line cuts off upon the axis of X exactly the segment desired, $\frac{y_4}{y_1}$. For the line a (Fig. 8) has the equation

$$(y_4 - 4)\, y + 2x = 2y_4.$$

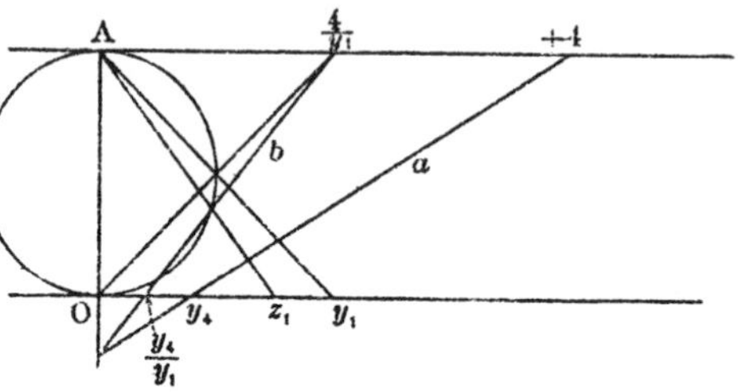

Fig. 8.

It cuts off upon the vertical diameter the segment $\frac{2y_4}{y_4 - 4}$. The equation of the line b is then

$$2y_1 x + (y_4 - 4)\, y = 2y_4,$$

and its intersection with the axis of X has the abscissa $\frac{y_4}{y_1}$.

If we project the upper intersection of the line b with the circle from A upon the axis of X, we obtain $z_1 = 2\cos\frac{2\pi}{17}$. If we desire the simple cosine itself we have only to draw a diameter parallel to the axis of X, on which our last projecting ray cuts off directly $\cos\frac{2\pi}{17}$. A perpendicular erected at this point gives immediately the first and sixteenth vertices of the regular polygon of 17 sides.

The period z_1 was chosen arbitrarily; we might construct in the same way every other period of two terms and so find the remaining cosines. These constructions, made on separate figures so as to be followed more easily, have been combined in a single figure (Fig. 9), which gives the complete construction of the regular polygon of 17 sides.

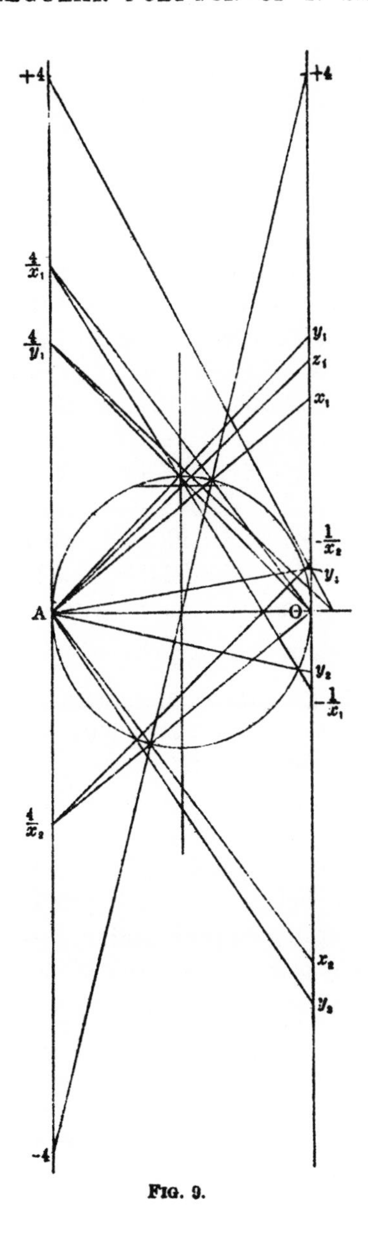

+4
+4
$\frac{4}{x_1}$
$\frac{4}{y_1}$
y_1
z_1
x_1
$-\frac{1}{x_2}$
y_2
A
O
y_2
$-\frac{1}{x_1}$
$\frac{4}{x_2}$
x_2
y_2
-4

FIG. 9.

CHAPTER V.

General Considerations on Algebraic Constructions.

1. We shall now lay aside the matter of construction with straight edge and compasses. Before quitting the subject we may mention a new and very simple method of effecting certain constructions, *paper folding.* Hermann Wiener* has shown how by paper folding we may obtain the network of the regular polyhedra. Singularly, about the same time a Hindu mathematician, Sundara Row, of Madras, published a little book, *Geometrical Exercises in Paper Folding* (Madras, Addison & Co., 1893), in which the same idea is considerably developed. The author shows how by paper folding we may construct by points such curves as the ellipse, cissoid, etc.

2. Let us now inquire how to solve geometrically problems whose analytic form is an equation of the third or of higher degree, and in particular, let us see how the ancients succeeded. The most natural method is by means of the conics, of which the ancients made much use. For example, they found that by means of these curves they were enabled to solve the problems of the duplication of the cube and the trisection of the angle. We shall in this place give only a general sketch of the process, making use of the language of modern mathematics for greater simplicity.

Let it be required, for instance, to solve graphically the cubic equation

$$x^3 + ax^2 + bx + c = 0,$$

or the biquadratic,

$$x^4 + ax^3 + bx^2 + cx + d = 0.$$

* See Dyck, *Katalog der Münchener mathematischen Ausstellung von 1893*, Nachtrag, p. 52.

Put $x^2 = y$; our equations become

$$xy + ay + bx + c = 0$$

and

$$y^2 + axy + by + cx + d = 0.$$

The roots of the equations proposed are thus the abscissas of the points of intersection of the two conics.

The equation

$$x^2 = y$$

represents a parabola with axis vertical. The second equation,

$$xy + ay + bx + c = 0,$$

represents an hyperbola whose asymptotes are parallel to the axes of reference (Fig. 10). One of the four points of inter-

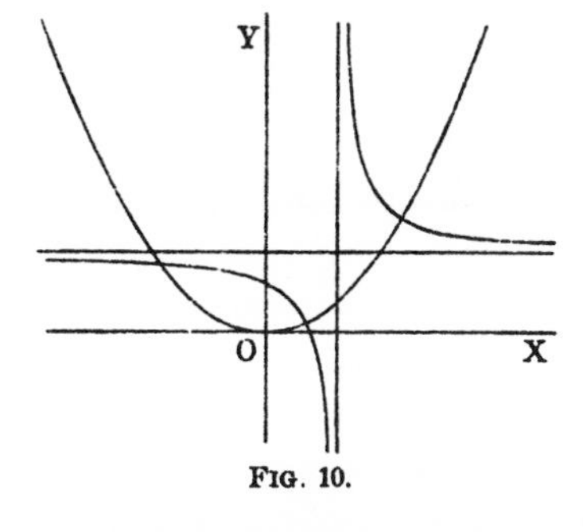

FIG. 10.

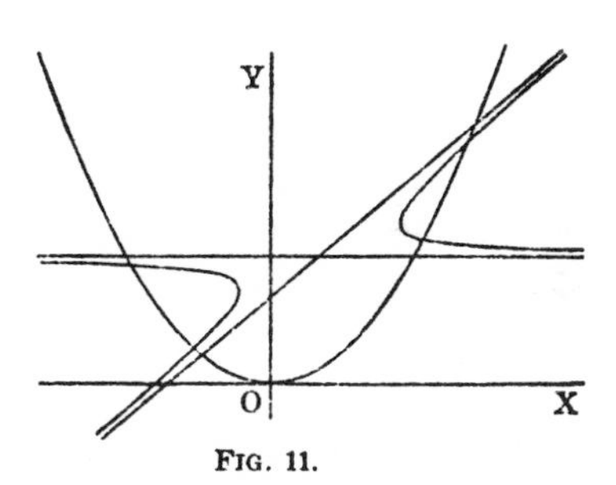

FIG. 11.

section is at infinity upon the axis of Y, the other three at a finite distance, and their abscissas are the roots of the equation of the third degree.

In the second case the parabola is the same. The hyperbola (Fig. 11) has again one asymptote parallel to the axis of X while the other is no longer perpendicular to this axis. The curves now have four points of intersection at a finite distance.

The methods of the ancient mathematicians are given in detail in the elaborate work of M. Cantor, *Geschichte der Mathematik* (Leipzig, 1894, 2d ed.). Especially interesting is Zeuthen, *Die Kegelschnitte im Altertum* (Kopenhagen, 1886, in German edition). As a general compendium we may mention Baltzer, *Analytische Geometrie* (Leipzig, 1882).

3. Beside the conics, the ancients used for the solution of the above-mentioned problems, higher curves constructed for this very purpose. We shall mention here only the *Cissoid* and the *Conchoid.*

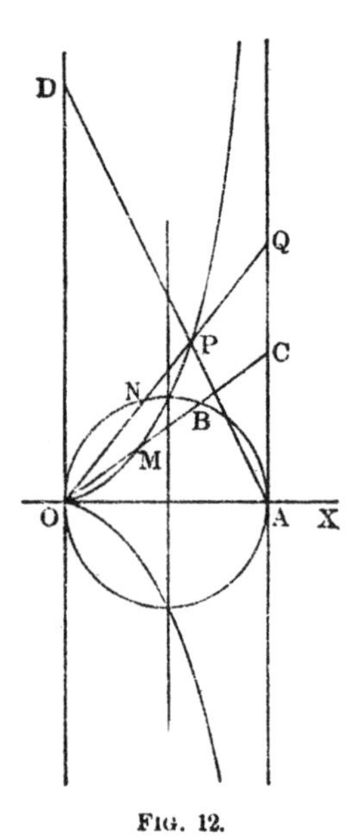

FIG. 12.

The *cissoid of Diocles* (c. 150 B.C.) may be constructed as follows (Fig. 12): To a circle draw a tangent (in the figure the vertical tangent on the right) and the diameter perpendicular to it. Draw lines from O, the vertex of the circle thus determined, to points upon the tangent, and lay off from O upon each the segment lying between its intersection with the circle and the tangent. The locus of points so determined is the *cissoid.*

To derive the equation, let r be the radius vector, θ the angle it makes with the axis of X. If we produce r to the tangent on the right, and call the diameter of the circle 1, the total segment equals $\frac{1}{\cos\theta}$. The portion cut off by the circle is $\cos\theta$. The difference of the two segments is r, and hence

$$r = \frac{1}{\cos\theta} - \cos\theta = \frac{\sin^2\theta}{\cos\theta}.$$

By transformation of coördinates we obtain the Cartesian equation,

$$(x^2+y^2)\,x-y^2=0.$$

The curve is of the third order, has a cusp at the origin, and is symmetric to the axis of X. The vertical tangent to the circle with which we began our construction is an asymptote. Finally the cissoid cuts the line at infinity in the circular points.

To show how to solve the Delian problem by the use of this curve, we write its equation in the following form:

$$\left(\frac{y}{x}\right)^3=\frac{y}{1-x}.$$

We now construct the straight line,

$$\frac{y}{x}=\lambda.$$

This cuts off upon the tangent $x=1$ the segment λ, and intersects the cissoid in a point for which

$$\frac{y}{1-x}=\lambda^3.$$

This is the equation of a straight line passing through the point $y=0$, $x=1$, and hence of the line joining this point to the point of the cissoid.

This line cuts off upon the axis of Y the intercept λ^3.

We now see how $\sqrt[3]{2}$ may be constructed. Lay off upon the axis of Y the intercept 2, join this point to the point $x=1$, $y=0$, and through its intersection with the cissoid draw a line from the origin to the tangent $x=1$. The intercept on this tangent equals $\sqrt[3]{2}$.

4. The *conchoid of Nicomedes* (*c.* 150 B.C.) is constructed as follows: Let O be a fixed point, a its distance from a fixed line. If we pass a pencil of rays through O and lay off on each ray from its intersection with the fixed line in both directions a segment b, the locus of the points so determined is the *conchoid.* According as b is greater or less than a, the origin is a node or a conjugate point; for $b = a$ it is a cusp (Fig. 13).

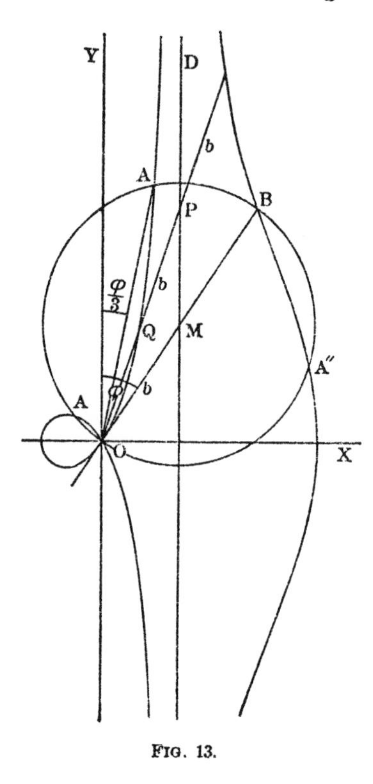

FIG. 13.

Taking for axes of X and Y the perpendicular and parallel through O to the fixed line, we have

$$\frac{r}{x} = \frac{b}{x - a};$$

whence

$$(x^2 + y^2)(x - a)^2 - b^2x^2 = 0.$$

The conchoid is then of the fourth order, has a double point at the origin, and is composed of two branches having for common asymptote the line $x = a$. Further, the factor $(x^2 + y^2)$ shows that the curve passes through the circular points at infinity, a matter of immediate importance.

We may trisect any angle by means of this curve in the following manner: Let $\phi = MOY$ (Fig. 13) be the angle to be divided into three equal parts. On the side OM lay off $OM = b$, an arbitrary length. With M as a center and radius b describe a circle, and through M perpendicular to the axis of X with origin O draw a vertical line representing the asymptote of the conchoid to be constructed. Construct the conchoid. Connect O with A, the intersection of the circle and the conchoid. Then is $\angle$ AOY one third of $\angle \phi$, as is easily seen from the figure.

Our previous investigations have shown us that the problem of the trisection of the angle is a problem of the third degree. It admits the three solutions

$$\frac{\phi}{3}, \quad \frac{\phi+2\pi}{3}, \quad \frac{\phi+4\pi}{3}.$$

Every algebraic construction which solves this problem by the aid of a curve of higher degree must obviously furnish all the solutions. Otherwise the equation of the problem would not be irreducible. These different solutions are shown in the figure. The circle and the conchoid intersect in eight points. Two of them coincide with the origin, two others with the circular points at infinity. None of these can give a solution of the problem. There remain, then, four points of intersection, so that we seem to have one too many. This is due to the fact that among the four points we necessarily find the point B such that OMB $= 2$ b, a point which may be determined without the aid of the curve. There actually remain then only three points corresponding to the three roots furnished by the algebraic solution.

5. In all these constructions with the aid of higher algebraic curves, we must consider the practical execution. We need an instrument which shall trace the curve by a continuous movement, for a construction by points is simply a method of approximation. Several instruments of this sort have been constructed; some were known to the ancients. Nicomedes invented a simple device for tracing the conchoid. It is the oldest of the kind besides the straight edge and compasses. (Cantor, I, p. 302.) A list of instruments of more recent construction may be found in Dyck's Katalog, pp. 227–230. 340. and Nachtrag, pp. 42, 43.

PART II

TRANSCENDENTAL NUMBERS AND THE QUADRATURE OF THE CIRCLE

PART II

TRANSCENDENTAL NUMBERS AND THE QUADRATURE OF THE CIRCLE

CHAPTER I.

Cantor's Demonstration of the Existence of Transcendental Numbers.

1. Let us represent numbers as usual by points upon the axis of abscissas. If we restrict ourselves to rational numbers the corresponding points will fill the axis of abscissas densely throughout (*überall dicht*), *i.e.*, in any interval no matter how small there is an infinite number of such points. Nevertheless, as the ancients had already discovered, the continuum of points upon the axis is not exhausted in this way; between the rational numbers come in the irrational numbers, and the question arises whether there are not distinctions to be made among the irrational numbers.

Let us define first what we mean by *algebraic numbers.* Every root of an algebraic equation

$$a_0\omega^n + a_1\omega^{n-1} + \cdots + a_{n-1}\omega + a_n = 0$$

with integral coefficients is called an algebraic number. Of course we consider only the real roots. Rational numbers occur as a special case in equations of the form

$$a_0\omega + a_1 = 0.$$

We now ask the question: Does the totality of real algebraic numbers form a continuum, or a discrete series such that other numbers may be inserted in the intervals? These new numbers, the so-called *transcendental* numbers, would then be characterized by this property, that they cannot be roots of an algebraic equation with integral coefficients.

This question was answered first by Liouville (*Comptes rendus*, 1844, and Liouville's *Journal*, Vol. XVI, 1851), and in fact the existence of transcendental numbers was demonstrated by him. But his demonstration, which rests upon the theory of continued fractions, is rather complicated. The investigation is notably simplified by using the developments given by Georg Cantor in a memoir of fundamental importance, *Ueber eine Eigenschaft des Inbegriffes reeller algebraischer Zahlen* (Crelle's *Journal*, Vol. LXXVII, 1873). We shall give his demonstration, making use of a more simple notion which Cantor, under a different form, it is true, suggested at the meeting of naturalists in Halle, 1891.

2. The demonstration rests upon the fact that algebraic numbers form a *countable* mass, while transcendental numbers do not. By this Cantor means that the former can be arranged in a certain order so that each of them occupies a definite place, is numbered, so to speak. This proposition may be stated as follows:

The manifoldness of real algebraic numbers and the manifoldness of positive integers can be brought into a one-to-one correspondence.

We seem here to meet a contradiction. The positive integers form only a portion of the algebraic numbers; since each number of the first can be associated with one and one only of the second, the part would be equal to the whole. This objection rests upon a false analogy. The proposition that the part is always less than the whole is not true for infinite masses. It is evident, for example, that we may establish a one-to-one correspondence between the aggregate of positive integers and the aggregate of positive even numbers, thus:

$$\begin{array}{ccccccc} 0 & 1 & 2 & 3 & \cdots & n & \cdots \\ 0 & 2 & 4 & 6 & \cdots & 2n & \cdots \end{array}$$

In dealing with infinite masses, the words *great* and *small* are inappropriate. As a substitute, Cantor has introduced the word *power* (*Mächtigkeit*), and says: *Two infinite masses have the same power when they can be brought into a one-to-one correspondence with each other.* The theorem which we have to prove then takes the following form: *The aggregrate of real algebraic numbers has the same power as the aggregate of positive integers.*

We obtain the aggregate of real algebraic numbers by seeking the real roots of all algebraic equations of the form

$$a_0\omega^n + a_1\omega^{n-1} + \cdots + a_{n-1}\omega + a_n = 0;$$

all the a's are supposed prime to one another, a_0 positive, and the equation irreducible. To arrange the numbers thus obtained in a definite order, we consider their *height* N as defined by

$$N = n - 1 + |a_0| + |a_1| + \cdots + |a_n|,$$

$|a_i|$ representing the absolute value of a_i, as usual. To a given number N corresponds a finite number of algebraic equations. For, N being given, the number n has certainly an upper limit, since N is equal to $n-1$ increased by positive numbers; moreover, the difference $N-(n-1)$ is a sum of positive numbers prime to one another, whose number is obviously finite.

N	n	$\|a_0\|$	$\|a_1\|$	$\|a_2\|$	$\|a_3\|$	$\|a_4\|$	EQUATION.	$\phi(N)$	ROOTS.
1	1	1	0				$x=0$	1	0
	2	0	0	0			—		
2	1	2	0				—	2	-1
		1	1				$x \pm 1 = 0$		$+1$
	2	1	0	0			—		
3	1	3	0				—	4	-2
		2	1				$2x \pm 1 = 0$		$-\frac{1}{2}$
		1	2				$x \pm 2 = 0$		$+\frac{1}{2}$
	2	2	0	0			—		$+2$
		1	1	0			—		
		1	0	1			—		
	3	1	0	0	0		—		
4	1	4	0				—	12	-3
		3	1				$3x \pm 1 = 0$		-1.61803
		2	2				—		-1.41421
		1	3				$x \pm 3 = 0$		-0.70711
	2	3	0	0			—		-0.61803
		2	1	0			—		-0.33333
		2	0	1			$2x^2 - 1 = 0$		$+0.33333$
		1	2	0			—		$+0.61803$
		1	1	1			$x^2 \pm x - 1 = 0$		$+0.70711$
		1	0	2			$x^2 - 2 = 0$		$+1.41421$
	3	2	0	0	0		—		$+1.61803$
		1	1	0	0		—		$+3$
		1	0	1	0		—		
		1	0	0	1		—		
	4	1	0	0	0	0	—		

Among these equations we must discard those that are reducible, which presents no theoretical difficulty. Since the number of equations corresponding to a given value of

N is limited, there corresponds to a determinate N only a finite mass of algebraic numbers. We shall designate this by $\phi(N)$. The table contains the values of $\phi(1)$, $\phi(2)$, $\phi(3)$, $\phi(4)$, and the corresponding algebraic numbers ω.

We arrange now the algebraic numbers according to their height, N, and the numbers corresponding to a single value of N in increasing magnitude. We thus obtain all the algebraic numbers, each in a determinate place. This is done in the last column of the accompanying table. It is, therefore, evident that algebraic numbers can be counted.

3. We now state the general proposition:

In any portion of the axis of abscissas, however small, there is an infinite number of points which certainly do not belong to a given countable mass.

Or, in other words:

The continuum of numerical values represented by a portion of the axis of abscissas, however small, has a greater power than any given countable mass.

This amounts to affirming the existence of transcendental numbers. It is sufficient to take as the countable mass the aggregate of algebraic numbers.

To demonstrate this theorem we prepare a table of algebraic numbers as before and write in it all the numbers in the form of decimal fractions. None of these will end in an infinite series of 9's. For the equality

$$1 = 0.999 \cdots 9 \cdots$$

shows that such a number is an exact decimal. If now we can construct a decimal fraction which is not found in our table and does not end in an infinite series of 9's it will certainly be a transcendental number. By means of a very simple process indicated by Georg Cantor we can find not only one but infinitely many transcendental numbers, even when the domain in which the number is to lie is very small Suppose, for example, that the first five decimals of the number are given. Cantor's process is as follows.

Take for 6th decimal a number different from 9 and from the 6th decimal of the *first algebraic* number, for 7th decimal a number different from 9 and from the 7th decimal of the *second algebraic* number, etc. In this way we obtain a decimal fraction which will not end in an infinite series of 9's and is certainly not contained in our table. The proposition is then demonstrated.

We see by this that (if the expression is allowable) there are far more transcendental numbers than algebraic. For when we determine the unknown decimals, avoiding the 9's, we have a choice among eight different numbers; we can thus form, so to speak, 8^{∞} transcendental numbers, even when the domain in which they are to lie is as small as we please.

CHAPTER II.

Historical Survey of the Attempts at the Computation and Construction of π.

In the next chapter we shall prove that the number π belongs to the class of transcendental numbers whose existence was shown in the preceding chapter. The proof was first given by Lindemann in 1882, and thus a problem was definitely settled which, so far as our knowledge goes, has occupied the attention of mathematicians for nearly 4000 years, the problem of the quadrature of the circle.

For, if the number π is not algebraic, it certainly cannot be constructed by means of straight edge and compasses. *The quadrature of the circle in the sense understood by the ancients is then impossible.* It is extremely interesting to follow the fortunes of this problem in the various epochs of science, as ever new attempts were made to find a solution with straight edge and compasses, and to see how these necessarily fruitless efforts worked for advancement in the manifold realm of mathematics.

The following brief historical survey is based upon the excellent work of Rudio: *Archimedes, Huygens, Lambert, Legendre, Vier Abhandlungen über die Kreismessung*, Leipzig, 1892. This book contains a German translation of the investigations of the authors named. While the mode of presentation does not touch upon the modern methods here discussed, the book includes many interesting details which are of practical value in elementary teaching.

1. Among the attempts to determine the ratio of the diameter to the circumference we may first distinguish the *empirical stage*, in which the desired end was to be attained by measurement or by direct estimation.

One of the oldest known mathematical documents, the Rhind Papyrus (*c.* 1650 B.C.), contains the problem in the well-known form, to transform a circle into a square of equal area. The writer of the papyrus lays down the following rule: Cut off $\frac{1}{9}$ of a diameter and construct a square upon the remainder; this has the same area as the circle. The value of π thus obtained is $(\frac{16}{9})^2 = 3.16 \cdots$, not very inaccurate. Much less accurate is the value $\pi = 3$, used in the Bible (1 Kings, 7. 23, 2 Chronicles, 4. 2).

2. The Greeks rose above this empirical standpoint, and especially Archimedes, who, in his work κύκλου μέτρησις, computed the area of the circle by the aid of inscribed and circumscribed polygons, as is still done in the schools. His method remained in use till the invention of the differential calculus; it was especially developed and rendered practical by Huygens (d. 1654) in his work, *De circuli magnitudine inventa.*

As in the case of the duplication of the cube and the trisection of the angle the Greeks sought also to effect the quadrature of the circle by the help of higher curves.

Consider for example the curve $y = \sin^{-1} x$, which represents the sinusoid with axis vertical. Geometrically, π appears as a particular ordinate of this curve; from the standpoint of the theory of functions, as a particular value of our transcendental function. Any apparatus which describes a transcendental curve we shall call a transcendental apparatus. A transcendental apparatus which traces the sinusoid gives us a geometric construction of π.

In modern language the curve $y = \sin^{-1} x$ is called an

integral curve because it can be defined by means of the integral of an algebraic function,

$$y = \int_0^x \frac{dx}{\sqrt{1-x^2}}.$$

The ancients called such a curve a *quadratrix* or *τετραγωνίζουσα*. The best known is the *quadratrix of Dinostratus* (*c.* 350 B.C.) which, however, had already been constructed by Hippias of Elis (*c.* 420 B.C.) for the trisection of an angle. Geometrically it may be defined as follows. Having given a circle and two perpendicular radii OA and OB, two points M and L move with constant velocity, one upon the radius OB, the other upon the arc AB (Fig. 14). Starting at the same time at O and A, they arrive simultaneously at B. The point of intersection P of OL and the parallel to OA through M describes the quadratrix.

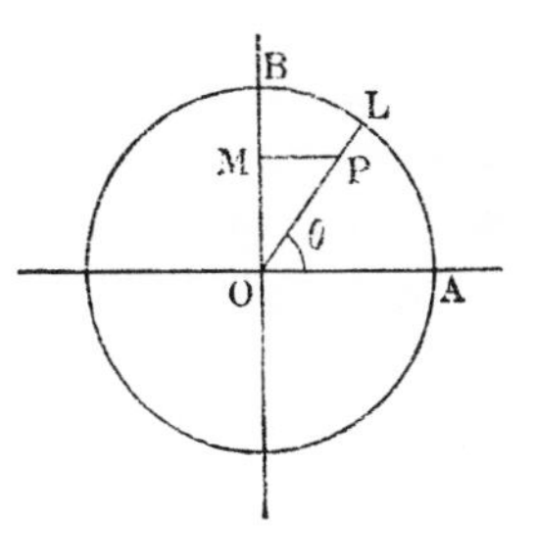

FIG. 14.

From this definition it follows that y is proportional to θ. Further, since for $y = 1$, $\theta = \frac{\pi}{2}$ we have

$$\theta = \frac{\pi}{2} y;$$

and from $\theta = \tan^{-1}\frac{y}{x}$ the equation of the curve becomes

$$\frac{y}{x} = \tan \frac{\pi}{2} y.$$

It meets the axis of X at the point whose abscissa is

$$x = \lim \frac{y}{\tan \frac{\pi}{2} y}, \text{ for } y = 0;$$

hence $$x = \frac{2}{\pi}.$$

According to this formula the radius of the circle is the mean proportional between the length of the quadrant and the abscissa of the intersection of the quadratrix with the axis of X. This curve can therefore be used for the rectification and hence also for the quadrature of the circle. This use of the quadratrix amounts, however, simply to a geometric formulation of the problem of rectification so long as we have no apparatus for describing the curve by continuous movement.

Fig. 15 gives an idea of the form of the curve with the branches obtained by taking values of θ greater than π or

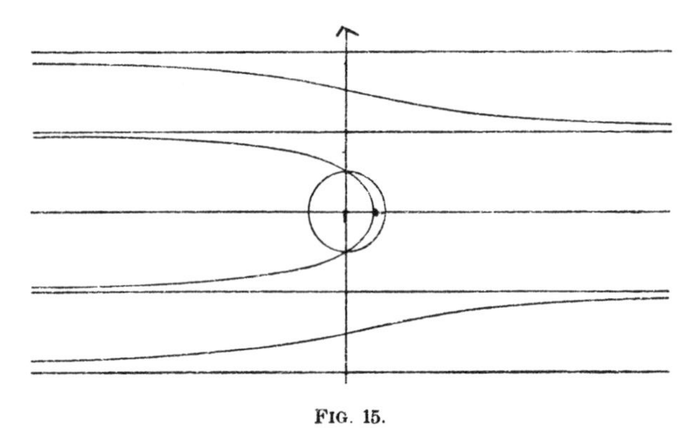

FIG. 15.

less than $-\pi$. Evidently the quadratrix of Dinostratus is not so convenient as the curve $y = \sin^{-1} x$, but it does not appear that the latter was used by the ancients.

3. The period from 1670 to 1770, characterized by the names of Leibnitz, Newton, and Euler, saw the rise of modern analysis. Great discoveries followed one another in such an almost unbroken series that, as was natural, critical rigor fell into the background. For our purposes the development of the theory of series is especially important. Numerous methods were deduced for approximating the value of π. It will suffice to mention the so-called *Leibnitz series* (known, however, before Leibnitz):

$$\frac{\pi}{4} = 1 - \frac{1}{3} + \frac{1}{5} - \frac{1}{7} \cdots$$

This same period brings the discovery of the mutual dependence of e and π. The number e, natural logarithms, and hence the exponential function, are first found in principle in the works of Napier (1614). This number seemed at first to have no relation whatever to the circular functions and the number π until Euler had the courage to make use of imaginary exponents. In this way he arrived at the celebrated formula

$$e^{ix} = \cos x + i \sin x,$$

which, for $x = \pi$, becomes

$$e^{i\pi} = -1.$$

This formula is certainly one of the most remarkable in all mathematics. The modern proofs of the transcendence of π are all based on it, since the first step is always to show the transcendence of e.

4. After 1770 critical rigor gradually began to resume its rightful place. In this year appeared the work of Lambert: *Vorläufige Kenntnisse für die so die Quadratur des Cirkuls suchen.* Among other matters the irrationality of π is discussed. In 1794 Legendre, in his *Éléments de géométrie*, showed conclusively that π and π^2 are irrational numbers.

5. But a whole century elapsed before the question was investigated from the modern point of view. The starting-point was the work of Hermite: *Sur la fonction exponentielle* (*Comptes rendus*, 1873, published separately in 1874). The transcendence of e is here proved.

An analogous proof for π, closely related to that of Hermite, was given by Lindemann: *Ueber die Zahl* π (*Mathematische Annalen*, XX, 1882. See also the Proceedings of the Berlin and Paris academies).

The question was then settled for the first time, but the investigations of Hermite and Lindemann were still very complicated.

The first simplification was given by Weierstrass in the *Berliner Berichte* of 1885. The works previously mentioned were embodied by Bachmann in his text-book, *Vorlesungen über die Natur der Irrationalzahlen*, 1892.

But the spring of 1893 brought new and very important simplifications. In the first rank should be named the memoirs of Hilbert in the *Göttinger Nachrichten.* Still Hilbert's proof is not absolutely elementary: there remain traces of Hermite's reasoning in the use of the integral

$$\int_0^\infty z^\rho e^{-z} dz = \rho\,!$$

But Hurwitz and Gordan soon showed that this transcendental formula could be done away with (*Göttinger Nachrichten; Comptes rendus;* all three papers are reproduced with some extensions in *Mathematische Annalen*, Vol. XLIII).

The demonstration has now taken a form so elementary that it seems generally available. In substance we shall follow Gordan's mode of treatment.

CHAPTER III.

The Transcendence of the Number e.

1. We take as the starting-point for our investigation the well-known series

$$e^x = 1 + \frac{x}{1} + \frac{x^2}{2!} + \cdots \frac{x^n}{n!} + \cdots,$$

which is convergent for all finite values of x. The difference between practical and theoretical convergence should here be insisted on. Thus, for $x = 1000$ the calculation of e^{1000} by means of this series would obviously not be feasible. Still the series certainly converges theoretically; for we easily see that after the 1000th term the factorial $n!$ in the denominator increases more rapidly than the power which occurs in the numerator. This circumstance that $\frac{x^n}{n!}$ has for any finite value of x the limit zero when n becomes infinite has an important bearing upon our later demonstrations.

We now propose to establish the following proposition:

The number e *is not an algebraic number, i.e.*, an equation with integral coefficients of the form

$$F(e) = C_0 + C_1e + C_2e^2 + \cdots + C_ne^n = 0$$

is impossible. The coefficients C_i may be supposed prime to one another.

We shall use the indirect method of demonstration, showing that the assumption of the above equation leads to an absurdity. The absurdity may be shown in the following way. We multiply the members of the equation $F(e)=0$ by a certain integer M so that

$$MF(e)=MC_0+MC_1e+MC_2e^2+\cdots+MC_ne^n=0.$$

We shall show that the number M can be chosen so that

(1) Each of the products Me, Me^2, $\cdots Me^n$ may be separated into an entire part M_κ and a fractional part ϵ_κ, and our equation takes the form

$$MF(e)=MC_0+M_1C_1+M_2C_2+\cdots+M_nC_n$$
$$+C_1\epsilon_1+C_2\epsilon_2+\cdots+C_n\epsilon_n=0;$$

(2) The integral part

$$MC_0+M_1C_1+\cdots+M_nC_n$$

is not zero. This will result from the fact that when divided by a prime number it gives a remainder different from zero;

(3) The expression

$$C_1\epsilon_1+C_2\epsilon_2+\cdots+C_n\epsilon_n$$

can be made as small a fraction as we please.

These conditions being fulfilled, the equation assumed is manifestly impossible, since the sum of an integer different from zero, and a proper fraction, cannot equal zero.

The salient point of the proof may be stated, though not quite accurately, as follows:

With an exceedingly small error we may assume $e, e^2, \cdots e^n$ proportional to integers which certainly do not satisfy our assumed equation.

2. We shall make use in our proof of a symbol h^r and a certain polynomial $\phi(x)$.

The symbol h^r is simply another notation for the factorial $r!$ Thus, we shall write the series for e^x in the form

$$e^x=1+\frac{x}{h}+\frac{x^2}{h^2}+\cdots+\frac{x^n}{h^n}+\cdots$$

The symbol has no deeper meaning; it simply enables us to write in more compact form every formula containing powers and factorials.

Suppose, *e.g.*, we have given a developed polynomial

$$f(x) = \sum_r c_r x^r.$$

We represent by $f(h)$, and write under the form $\sum_r c_r h^r$, the sum

$$c_1 \cdot 1 + c_2 \cdot 2! + c_3 \cdot 3! + \cdots + c_n \cdot n!$$

But if $f(x)$ is not developed, then to calculate $f(h)$ is to develop this polynomial in powers of h and finally replace h^r by $r!$. Thus, for example,

$$f(k+h) = \sum_r c_r (k+h)^r = \sum_r c'_r \cdot h^r = \sum_r c'_r \cdot r!,$$

the c'_r depending on k.

The polynomial $\phi(x)$ which we need for our proof is the following remarkable expression

$$\phi(x) = x^{p-1} \frac{[(1-x)(2-x)\cdots(n-x)]^p}{(p-1)!},$$

where p is a prime number, n the degree of the algebraic equation assumed to be satisfied by e. We shall suppose p greater than n and $|C_0|$, and later we shall make it increase without limit.

To get a geometric picture of this polynomial $\phi(x)$ we construct the curve

$$y = \phi(x).$$

At the points $x = 1, 2, \cdots n$ the curve has the axis of X as an inflexional tangent, since it meets it in an odd number of points, while at the origin the axis of X is tangent without inflexion. For values of x between 0 and n the curve remains in the neighborhood of the axis of X; for greater values of x it recedes indefinitely.

Of the function $\phi(x)$ we will now establish three important properties :

1. x *being supposed given and* p *increasing without limit,* $\phi(x)$ *tends toward zero, as does also the sum of the absolute values of its terms.*

Put $u = x(1-x)(2-x)\cdots(n-x)$; we may then write

$$\phi(x) = \frac{u^{p-1}}{(p-1)!}\,\frac{u}{x},$$

which for p infinite tends toward zero.

To have the sum of the absolute values of $\phi(x)$ it is sufficient to replace $-x$ by $|x|$ in the undeveloped form of $\phi(x)$. The second part is then demonstrated like the first.

2. h *being an integer,* $\phi(h)$ *is an integer not divisible by* p *and therefore different from zero.*

Develop $\phi(x)$ in increasing powers of x, noticing that the terms of lowest and highest degree respectively are of degree $p-1$ and $np+p-1$. We have

$$\phi(x) = \sum_{r=p-1}^{r=np+p-1} c_r x^r = \frac{c' x^{p-1}}{(p-1)!} + \frac{c'' x^p}{(p-1)!} + \cdots \pm \frac{x^{np+p-1}}{(p-1)!}.$$

Hence

$$\phi(h) = \sum_{r=p-1}^{r=np+p-1} c_r h^r.$$

Leaving out of account the denominator $(p-1)!$, which occurs in all the terms, the coefficients c_r are integers. This denominator disappears as soon as we replace h^r by $r!$, since the factorial of least degree is $h^{p-1} = (p-1)!$. All the terms of the development after the first will contain the factor p. As to the first, it may be written

$$\frac{(1\cdot 2\cdot 3\cdots n)^p\cdot(p-1)!}{(p-1)!} = (n!)^p$$

and is certainly not divisible by p since $p > n$.

Therefore $\qquad \phi(h) \equiv (n!)^p \pmod{p}$,

and hence $\qquad \phi(h) \neq 0$.

Moreover, $\phi(h)$ is a very large number; even its last term alone is very large, viz.:

$$\frac{(np+p-1)!}{(p-1)!} = p(p+1)\cdots(np+p-1).$$

3. h *being an integer, and* k *one of the numbers* $1, 2 \cdots n$, $\phi(h+k)$ *is an integer divisible by* p.

We have $\phi(h+k) = \sum_r c_r (h+k)^r = \sum_r c'_r h^r,$

a formula in which we are to replace h^r by $r!$ only after having arranged the development in increasing powers of h.

According to the rules of the symbolic calculus, we have first

$$\phi(h+k) = (h+k)^{p-1} \frac{[(1-k-h)(2-k-h)\cdots(-h)\cdots(n-k-h)]^p}{(p-1)!}.$$

One of the factors in the brackets reduces to $-h$; hence the term of lowest degree in h in the development is of degree p. We may then write

$$\phi(h+k) = \sum_{r=p}^{r=np+p-1} c'_r h^r.$$

The coefficients still have for numerators integers and for denominator $(p-1)!$. As already explained, this denominator disappears when we replace h^r by $r!$. But now all the terms of the development are divisible by p; for the first may be written

$$\frac{(-1)^{kp}\cdot k^{p-1}[(k-1)!(n-k)!]^p\cdot p!}{(p-1)!} = (-1)^{kp}k^{p-1}[(k-1)!\cdot(n-k)!]^p\cdot p.$$

$\phi(h+k)$ is then divisible by p.

3. We can now show that the equation

$$F(e) = C_0 + C_1 e + C_2 e^2 + \cdots + C_n e^n = 0$$

is impossible.

For the number M, by which we multiply the members of this equation, we select $\phi(h)$, so that

$$\phi(h) F(e) = C_0\phi(h) + C_1\phi(h)e + C_2\phi(h)e^2 + \cdots + C_n\phi(h)e^n.$$

Let us try to decompose any term, such as $C_k\phi(h)e^k$, into an integer and a fraction. We have

$$e^k \cdot \phi(h) = e^k \sum_r c_r h^r.$$

Considering the series development of e^k, any term of this sum, omitting the constant coefficient, has the form

$$e^k \cdot h^r = h^r + \frac{h^r \cdot k}{1} + \frac{h^r \cdot k^2}{2!} + \cdots + \frac{h^r \cdot k^r}{r!} + \frac{h^r \cdot k^{r+1}}{(r+1)!} + \cdots$$

Replacing h^r by $r!$, or what amounts to the same thing, by one of the quantities

$$rh^{r-1},\ r(r-1)h^{r-2} \cdots,\ r(r-1) \cdots 3 \cdot h^2,\ r(r-1) \cdots 2 \cdot h,$$

and simplifying the successive fractions,

$$e^k \cdot h^r = h^r + \frac{r}{1} \cdot h^{r-1}k + \frac{r(r-1)}{2!} h^{r-2}k^2 + \cdots + \frac{r}{1} hk^{r-1} + k^r$$

$$+ k^r \left[\frac{k}{r+1} + \frac{k^2}{(r+1)(r+2)} + \cdots\right].$$

The first line has the same form as the development of $(h+k)^r$; in the parenthesis of the second line we have the series

$$0 + \frac{k}{r+1} + \frac{k^2}{(r+1)(r+2)} + \cdots$$

whose terms are respectively less than those of the series

$$e^k = 1 + k + \frac{k^2}{2!} + \frac{k^3}{3!} + \cdots$$

The second line in the expansion of $e^k \cdot h^r$ may therefore be represented by an expression of the form

$$q_{r,k} \quad e^k \cdot k^r,$$

$q_{r,k}$ being a proper fraction.

Effecting the same decomposition for each term of the sum

$$e^k \sum c_r h^r$$

it takes the form

$$e^k \sum_r c_r h^r = \sum_r c_r (h + k)^r + e^k \sum_r q_{r,k} c_r k^r.$$

The first part of this sum is simply $\phi(h + k)$; this is a number divisible by p (2, 3). Further (2, 1),

$$\phi(k) = \sum_r |c_r k^r|$$

tends toward zero when p becomes infinite: the same is true *a fortiori* of $\sum\limits_r q_{r,k} c_r k^r$, and also, since e^k is a finite quantity, of $e^k \sum\limits_r q_{r,k} c_r k^r$, which we may represent by ϵ_k.

The term under consideration, $C_k e^k \phi(h)$, has then been put under the form of an integer $C_k \phi(h + k)$ and a quantity $C_k \epsilon_k$ which, by a suitable choice of p, may be made as small as we please.

Proceeding similarly with all the terms, we get finally

$$F(e)\,\phi(h) = C_0 \phi(h) + C_1 \phi(h + 1) + \cdots + C_n \phi(h + n)$$
$$+ C_1 \epsilon_1 + C_2 \epsilon_2 + \cdots + C_n \epsilon_n.$$

It is now easy to complete the demonstration. All the terms of the first line after the first are divisible by p; for the first, $|C_0|$ is less than p; $\phi(h)$ is not divisible by p; hence $C_0 \phi(h)$ is not divisible by the prime number p. Consequently the sum of the numbers of the first line is not zero.

The numbers of the second line are finite in number; each of them can be made smaller than any given number by a suitable choice of p; and therefore the same is true of their sum.

Since an integer not zero and a fraction cannot have zero for a sum, the assumed equation is impossible.

Thus, the transcendence of e, or Hermite's Theorem, is demonstrated.

CHAPTER IV.

The Transcendence of the Number π.

1. The demonstration of the transcendence of the number π given by Lindemann is an extension of Hermite's proof in the case of e. While Hermite shows that an integral equation of the form

$$C_0 + C_1 e + C_2 e^2 + \cdots + C_n e^n = 0$$

cannot exist, Lindemann generalizes this by introducing in place of the powers $e, e^2 \cdots$ sums of the form

$$e^{k_1} + e^{k_2} + \cdots + e^{k_N}$$
$$e^{l_1} + e^{l_2} + \cdots + e^{l_N}$$
$$\cdot \quad \cdot \quad \cdot \quad \cdot \quad \cdot \quad \cdot \quad \cdot$$

where the k's are associated algebraic numbers, *i.e.*, roots of an algebraic equation, with integral coefficients, of the degree N; the l's roots of an equation of degree N′, etc. Moreover, some or all of these roots may be imaginary.

Lindemann's general theorem may be stated as follows:

The number e *cannot satisfy an equation of the form*

$$(1) \quad C_0 + C_1(e^{k_1} + e^{k_2} + \cdots + e^{k_N}) + C_2(e^{l_1} + e^{l_2} + \cdots + e^{l_{N'}}) + \cdots = 0$$

where the coefficients C_i *are integers and the exponents* $k_i, l_i, \cdots$ *are respectively associated algebraic numbers.*

The theorem may also be stated:

The number e *is not only not an algebraic number and therefore a transcendental number simply, but it is also not an interscendental* number and therefore a transcendental number of higher order.*

* Leibnitz calls a function x^λ, where λ is an algebraic irrational, an interscendental function.

Let

$$ax^N + a_1x^{N-1} + \cdots + a_N = 0$$

be the equation having for roots the exponents k_i;

$$bx^{N'} + b_1x^{N'-1} + \cdots + b_{N'} = 0$$

that having for roots the exponents l_i, etc. These equations are not necessarily irreducible, nor the coefficients of the first terms equal to 1. It follows that the symmetric functions of the roots which alone occur in our later developments need not be integers.

In order to obtain integral numbers it will be sufficient to consider symmetric functions of the quantities

$$ak_1, ak_2, \cdots ak_N,$$
$$bl_1, bl_2, \cdots bl_{N'}, \text{ etc.}$$

These numbers are roots of the equations

$$y^N + a_1y^{N-1} + a_2ay^{N-2} + \cdots + a_Na^{N-1} = 0,$$
$$y^{N'} + b_1y^{N'-1} + b_2by^{N'-2} + \cdots + b_{N'}b^{N'-1} = 0, \text{ etc.}$$

These quantities are integral associated algebraic numbers, and their rational symmetric functions real integers.

We shall now follow the same course as in the demonstration of Hermite's theorem.

We assume equation (1) to be true; we multiply both members by an integer M; and we decompose each sum, such as

$$M(e^{k_1} + e^{k_2} + \cdots + e^{k_N}),$$

into an integral part and a fraction, thus

$$M(e^{k_1} + e^{k_2} + \cdots + e^{k_N}) = M_1 + \epsilon_1,$$
$$M(e^{l_1} + e^{l_2} + \cdots + e^{l_{N'}}) = M_2 + \epsilon_2,$$
$$\cdots\cdots\cdots$$

Our equation then becomes

$$C_0M + C_1M_1 + C_2M_2 + \cdots$$
$$+ C_1\epsilon_1 + C_2\epsilon_2 + \cdots = 0.$$

We shall show that with a suitable choice of M the sum of the quantities in the first line represents an integer not divisible by a certain prime number p, and consequently different from zero; that the fractional part can be made as small as we please, and thus we come upon the same contradiction as before.

2. We shall again use the symbol $h^r = r!$ and select as the multiplier the quantity $M = \psi(h)$, where $\psi(x)$ is a generalization of $\phi(x)$ used in the preceding chapter, formed as follows:

$$\psi(x) = \frac{x^{p-1}}{(p-1)!}[(k_1 - x)(k_2 - x)\cdots(k_N - x)]^p \cdot a^{Np} \cdot a^{N'p} \cdot a^{N''p} \cdots$$
$$\cdot [(l_1 - x)(l_2 - x)\cdots(l_{N'} - x)]^p \cdot b^{Np} \cdot b^{N'p} \cdot b^{N''p} \cdots$$
$$\cdot \quad \cdot \quad \cdot \quad \cdot \quad \cdot \quad \cdot \quad \cdot \quad \cdot$$
$$\cdot \quad \cdot \quad \cdot \quad \cdot \quad \cdot \quad \cdot \quad \cdot \quad \cdot$$

where p is a prime number greater than the absolute value of each of the numbers

$$C_0, a, b, \cdots, a_N, b_{N'}, \cdots$$

and later will be assumed to increase without limit. As to the factors a^{Np}, $b^{N'p}$, $\cdots$, they have been introduced so as to have in the development of $\psi(x)$ symmetric functions of the quantities

$$ak_1, ak_2, \cdots, ak_N,$$
$$bl_1, bl_2, \cdots, bl_{N'},$$
$$\cdot \quad \cdot \quad \cdot \quad \cdot \quad \cdot$$

that is, rational integral numbers. Later on we shall have to develop the expressions

$$\sum_\nu \psi(k_\nu + h), \quad \sum_\nu \psi(l_\nu + h), \cdots$$

The presence of these same factors will still be necessary if we wish the coefficients of these developments to be integers each divided by $(p-1)!$.

1. $\psi(h)$ *is an integral number, not divisible by* p *and consequently different from zero.*

Arranging $\psi(h)$ in increasing powers of h, it takes the form

$$\psi(h) = \sum_{r=p-1}^{r=Np+N'p+\cdots+p-1} c_r h^r.$$

In this development all the coefficients have integral numerators and the common denominator $(p-1)!$.

The coefficient of the first term h^{p-1} may be written

$$\frac{1}{(p-1)!}\begin{array}{l}(ak_1 \cdot ak_2 \cdots ak_N)^p a^{N'p} a^{N''p} \cdots \\ (bl_1 \cdot bl_2 \cdots bl_{N'})^p b^{Np} b^{N''p} \cdots \\ \cdots\cdots\cdots\cdots\cdots\end{array}$$

$$= \frac{1}{(p-1)!}(-1)^{Np+N'p+\cdots}(a_N a^{N-1})^p a^{N'p} a^{N''p} \cdots (b_{N'} b^{N'-1})^p b^{Np} b^{N''p} \cdots$$

If in this term we replace h^{p-1} by its value $(p-1)!$ the denominator disappears. According to the hypotheses made regarding the prime number p, no factor of the product is divisible by p and hence the product is not.

The second term $c_p h^p$ becomes likewise an integer when we replace h^p by $p!$ but the factor p remains, and so for all of the following terms. Hence $\psi(h)$ is an integer not divisible by p.

2. *For* x, *a given finite quantity, and* p *increasing without limit,* $\psi(x) = \sum_r c_r x^r$ *tends toward zero, as does also the sum* $\sum_r |c_r x^r|$.

We may write

$$\psi(x) = \sum c_r x^r$$

$$= \frac{x^{p-1}}{(p-1)!}[a^N a^{N} \cdots b^N b^{N'} (k_1 - x)(k_2 - x) \cdots (k_N - x)$$

$$(l_1 - x)(l_2 - x) \cdots (l_{N'} - x) \cdots]^p.$$

Since for x of given value the expression in brackets is a constant, we may replace it by K. We then have

$$\psi(x) = \frac{(xK)^{p-1}}{(p-1)!} K,$$

a quantity which tends toward zero as p increases indefinitely.

The same reasoning will apply when each term of $\psi(x)$ is replaced by its absolute value.

3. *The expression* $\sum_{\nu=1}^{\nu=N} \psi(k_\nu + h)$ *is an integer divisible by* p.

We have

$$\psi(k_\nu + h) = \frac{a^p (k_\nu + h)^{p-1}}{(p-1)!} b^{Np} b^{N''p} \cdot \cdot \cdot$$
$$\cdot\, a^{(N-1)p} [(k_1 - k_\nu - h)(k_2 - k_\nu - h) \cdot \cdot \cdot (-h) \cdot \cdot \cdot (k_N - k_\nu - h)]^p$$
$$\cdot\, a^{N'p} b^{N'p} [(l_1 - k_\nu - h)(l_2 - k_\nu - h) \cdot \cdot \cdot (l_{N'} - k_\nu - h)]^p$$
$$\cdot \quad \cdot \quad \cdot \quad \cdot \quad \cdot \quad \cdot \quad \cdot \quad \cdot \quad \cdot \quad \cdot \quad \cdot \quad \cdot \quad \cdot$$

The νth factor of the expression in brackets in the second line is $-h$, and hence the term of lowest degree in h is h^p.

Consequently

$$\psi(k_\nu + h) = \sum_{r=p}^{r=Np+N'p+\cdots+p-1} c'_r h^r,$$

whence

$$\sum_{\nu=1}^{\nu=N} \psi(k_\nu + h) = \sum_{r=p}^{r=Np+N'p+\cdots+p-1} C'_r h^r.$$

The numerators of the coefficients C'_r are rational and integral, for they are integral symmetric functions of the quantities

$$\begin{array}{llll} ak_1, & ak_2, & \cdot \cdot \cdot, & ak_N, \\ bl_1, & bl_2, & \cdot \cdot \cdot, & bl_{N'}, \\ \cdot \quad \cdot \quad \cdot & \cdot \quad \cdot \quad \cdot & \cdot \quad \cdot \quad \cdot & \cdot \quad \cdot \quad \cdot \end{array}$$

and their common denominator is $(p-1)!$.

If we replace h^r by $r!$ the denominator disappears from all the coefficients, the factor p remains in every term, and hence the sum is an integer divisible by p.

Similarly for

$$\sum_{\nu=1}^{\nu=N} \psi(l_\nu + h) \cdot \cdot \cdot$$

We have thus established three properties of $\psi(x)$ analogous to those demonstrated for $\phi(x)$ in connection with Hermite's theorem.

3. We now return to our demonstration that the assumed equation

(1) $C_0+C_1(e^{k_1}+e^{k_2}+\cdots+e^{k_N})+C_2(e^{l_1}+e^{l_2}+\cdots e^{l_{N'}})+\cdots=0$

cannot be true. For this purpose we multiply both members by $\psi(h)$, thus obtaining

$$C_0\psi(h)+C_1[e^{k_1}\psi(h)+e^{k_2}\psi(h)+\cdots+e^{k_N}\psi(h)]+\cdots=0,$$

and try to decompose each of the expressions in brackets into a whole number and a fraction. The operation will be a little longer than before, for k may be a complex number of the form $k=k'+ik''$. We shall need to introduce $|k|=+\sqrt{k'^2+k''^2}$.

One term of the above sum is

$$e^k\cdot\psi(h)=e^k\sum_r c_rh^r=\sum_r c_r\cdot e^k\cdot h^r.$$

The product $e^k\cdot h^r$ may be written, as shown before,

$$e^k\cdot h^r=(h+k)^r+k^r\left[\frac{k}{r+1}+\frac{k^2}{(r+1)(r+2)}+\cdots\right]$$

The absolute value of every term of the series.

$$0+\frac{k}{r+1}+\frac{k^2}{(r+1)(r+2)}+\cdots$$

is less than the absolute value of the corresponding term in the series

$$e^k=1+\frac{k}{1}+\frac{k^2}{2!}+\cdots$$

Hence
$$\left|\frac{k}{r+1}+\frac{k^2}{(r+1)(r+2)}+\cdots\right|<e^{|k|}$$

or
$$\frac{k}{r+1}+\frac{k^2}{(r+1)(r+2)}+\cdots=q_{r,k}e^{|k|},$$

$q_{r,k}$ being a complex quantity whose absolute value is less than 1.

We may then write

$$e^k\cdot\psi(h)=\sum_r c_re^kh^r=\sum_r c_r(h+k)^r+\sum_r c_rq_{r,k}k^re^{|k|}$$
$$=\psi(h+k)+\sum_r c_rq_{r,k}k^r\cdot e^{|k|}.$$

By giving k in succession the indices 1, 2, $\cdots$ N, and forming the sum the equation becomes

$$e^{k_1}\psi(h)+e^{k_2}\psi(h)+\cdots+e^{k_N}\psi(h)$$
$$=\sum_{\nu=1}^{\nu=N}\psi(k_\nu+h)+\sum_{\nu=1}^{\nu=N}\{e^{|k_\nu|}\sum_r c_r k_\nu^r q_{r,k_\nu}\}.$$

Proceeding similarly with all the other sums, our equation takes the form

$$(2)\quad C_0\psi(h)+C_1\sum_{\nu=1}^{\nu=N}\psi(k_\nu+h)+C_2\sum_{\nu=1}^{\nu=N'}\psi(l_\nu+h)+\cdots$$
$$+C_1\sum_{\nu=1}^{\nu=N}\sum_r e^{|k_\nu|}c_r k_\nu^r q_{r,k_\nu}+C_2\sum_{\nu=1}^{\nu=N'}e^{|l_\nu|}c'_r l_\nu^r q_{r,l_\nu}+\cdots=0.$$

By 2, 2 we can make $\sum_r|c_r k^r|$ as small as we please by taking p sufficiently great. Since $|q_{r,k}|<1$, this will be true *a fortiori* of

$$\sum_r c_r k^r q_{r,k}$$

and hence also of

$$\sum_{\nu=1}^{\nu=N}\sum_r c_r k_\nu^r q_{r,k} e^{|k_\nu|}.$$

Since the coefficients C are finite in value and in number, the sum which occurs in the second line of (2) can, by increasing p, be made as small as we please.

The numbers of the first line are, after the first, all divisible by p (3), but the first number, $C_0\psi(h)$, is not (1). Therefore the sum of the numbers in the first line is not divisible by p and hence is different from zero. The sum of an integer and a fraction cannot be zero. Hence equation (2) is impossible and consequently also equation (1).*

4. We now come to a proposition more general than the preceding, but whose demonstration is an immediate conse-

* The proof for the more general case where $C_0=0$ may be reduced to this by multiplication by a suitable factor, or may be obtained directly by a proper modification of $\psi(h)$.

quence of the latter. For this reason we shall call it Lindemann's corollary.

The number e *cannot satisfy an equation of the form*

$$(3) \quad C'_0 + C'_1 e^{k_1} + C'_2 e^{l_1} + \cdots = 0,$$

in which the coefficients are integers even when the exponents $k_1, l_1, \cdots$ *are unrelated algebraic numbers.*

To demonstrate this, let $k_2, k_3, \cdots, k_N$ be the other roots of the equation satisfied by k_1; similarly for $l_2, l_3, \cdots, l_{N'}$, etc. Form all the polynomials which may be deduced from (3) by replacing k_1 in succession by the associated roots $k_2, \cdots$, l_1 by the associated roots $l_2, \cdots$ Multiplying the expressions thus formed we have the product

$$\prod_{\alpha, \beta, \cdots} \{C'_0 + C'_1 e^{k_\alpha} + C'_2 e^{l_\beta} + \cdots\} \quad \begin{bmatrix} \alpha = 1, 2, \cdots, N \\ \beta = 1, 2, \cdots, N' \\ \cdot \quad \cdot \quad \cdot \quad \cdot \quad \cdot \quad \cdot \end{bmatrix}$$

$$= C_0 + C_1(e^{k_1} + e^{k_2} + \cdots + e^{k_N}) + C_2(e^{k_1+k_2} + e^{k_2+k_3} + \cdots)$$
$$+ C_3(e^{k_1+l_1} + e^{k_1+l_2} + \cdots) + \cdots$$

In each parenthesis the exponents are formed symmetrically from the quantities $k_i, l_i, \cdots$, and are therefore roots of an algebraic equation with integral coefficients. Our product comes under Lindemann's theorem; hence it cannot be zero. Consequently none of its factors can be zero and the corollary is demonstrated.

We may now deduce a still more general theorem.

The number e *cannot satisfy an equation of the form*

$$C_0^{(1)} + C_1^{(1)} e^k + C_2^{(1)} e^l + \cdots = 0$$

where the coefficients as well as the exponents are unrelated algebraic numbers.

For, let us form all the polynomials which we can deduce from the preceding when for each of the expressions $C^{(1)}_i$ we substitute one of the associated algebraic numbers

$$C_i^{(2)}, C_i^{(3)}, \cdots C_i^{(N)}.$$

If we multiply the polynomials thus formed together we get the product

$$\prod_{\alpha,\beta,\gamma,\cdots} \{C_0^{(\alpha)} + C_1^{(\beta)}e^k + C_2^{(\gamma)}e^l + \cdots\} \quad \begin{bmatrix} \alpha = 1, 2, \cdots, N_0 \\ \beta = 1, 2, \cdots, N_1 \\ \gamma = 1, 2, \cdots, N_2 \\ \cdots\cdots \end{bmatrix}$$

$$\begin{aligned} = C_0 &+ C_k e^k + C_l e^l + \cdots \\ &+ C_{k,k} e^{k+k} + C_{k,l} e^{k+l} + \cdots \\ &+ \cdots\cdots\cdots \\ &+ \cdots\cdots\cdots, \end{aligned}$$

where the coefficients C are integral symmetric functions of the quantities

$$\begin{matrix} C_0^{(1)}, & C_0^{(2)}, & \cdots, & C_0^{(N_0)}, \\ C_1^{(1)}, & C_1^{(2)}, & \cdots, & C_1^{(N_1)}, \\ \cdots & \cdots & \cdots & \cdots \\ \cdots & \cdots & \cdots & \cdots \end{matrix}$$

and hence are rational. By the previous proof such an expression cannot vanish, and we have accordingly Lindemann's corollary in its most general form :

The number e *cannot satisfy an equation of the form*

$$C_0 + C_1 e^k + C_2 e^l + \cdots = 0$$

where the exponents $k, l, \cdots$ *as well as the coefficients* $C_0, C_1, \cdots$ *are algebraic numbers.*

This may also be stated as follows:

In an equation of the form

$$C_0 + C_1 e^k + C_2 e^l + \cdots = 0$$

the exponents and coefficients cannot all be algebraic numbers.

5. From Lindemann's corollary we may deduce a number of interesting results. First, the transcendence of π is an immediate consequence. For consider the remarkable equation

$$1 + e^{i\pi} = 0.$$

The coefficients of this equation are algebraic; hence the exponent $i\pi$ is not. Therefore, π is transcendental.

6. Again consider the function $y = e^x$. We know that $1 = e^0$. This seems to be contrary to our theorems about the transcendence of e. This is not the case, however. We must notice that the case of the exponent 0 was implicitly excluded. For the exponent 0 the function $\psi(x)$ would lose its essential properties and obviously our conclusions would not hold.

Excluding then the special case ($x = 0$, $y = 1$), Lindemann's corollary shows that in the equation $y = e^x$ or $x = \log_e y$, y and x, *i.e.*, the number and its natural logarithm, cannot be algebraic simultaneously. To an algebraic value of x corresponds a transcendental value of y, and conversely. This is certainly a very remarkable property.

If we construct the curve $y = e^x$ and mark all the algebraic points of the plane, *i.e.*, all points whose coördinates are algebraic numbers, the curve passes among them without meeting a single one except the point $x = 0$, $y = 1$. The theorem still holds even when x and y take arbitrary complex values. The exponential curve is then transcendental in a far higher sense than ordinarily supposed.

7. A further consequence of Lindemann's corollary is the transcendence, in the same higher sense, of the function $y = \sin^{-1} x$ and similar functions.

The function $y = \sin^{-1} x$ is defined by the equation

$$2ix = e^{iy} - e^{-iy}.$$

We see, therefore, that here also x and y cannot be algebraic simultaneously, excluding, of course, the values $x = 0$, $y = 0$. We may then enunciate the proposition in geometric form:

The curve $y = \sin^{-1} x$, *like the curve* $y = e^x$, *passes through no algebraic point of the plane, except* $x = 0$, $y = 0$.

CHAPTER V.

The Integraph and the Geometric Construction of π.

1. Lindemann's theorem demonstrates the transcendence of π, and thus is shown the impossibility of solving the old problem of the quadrature of the circle, not only in the sense understood by the ancients but in a far more general manner. It is not only impossible to construct π with straight edge and compasses, but there is not even a curve of higher order defined by an integral algebraic equation for which π is the ordinate corresponding to a rational value of the abscissa. An actual construction of π can then be effected only by the aid of a transcendental curve. If such a construction is desired, we must use besides straight edge and compasses a "transcendental" apparatus which shall trace the curve by continuous motion.

2. Such an apparatus is the *integraph*, recently invented and described by a Russian engineer, Abdank-Abakanowicz, and constructed by Coradi of Zürich.

This instrument enables us to trace the *integral curve*

$$Y = F(x) = \int f(x)\,dx$$

when we have given the *differential curve*

$$y = f(x).$$

For this purpose, we move the linkwork of the integraph so that the *guiding point* follows the differential curve; the *tracing point* will then trace the integral curve. For a fuller description of this ingenious instrument we refer to the original memoir (in German, Teubner, 1889; in French, Gauthier-Villars, 1889).

We shall simply indicate the principles of its working. For any point (x, y) of the differential curve construct the auxiliary triangle having for vertices the points (x, y), (x, 0), (x — 1, 0); the hypotenuse of this right-angled triangle makes with the axis of X an angle whose tangent $= y$.

Hence, *this hypotenuse is parallel to the tangent to the integral curve at the point* (X, Y) *corresponding to the point* (x, y).

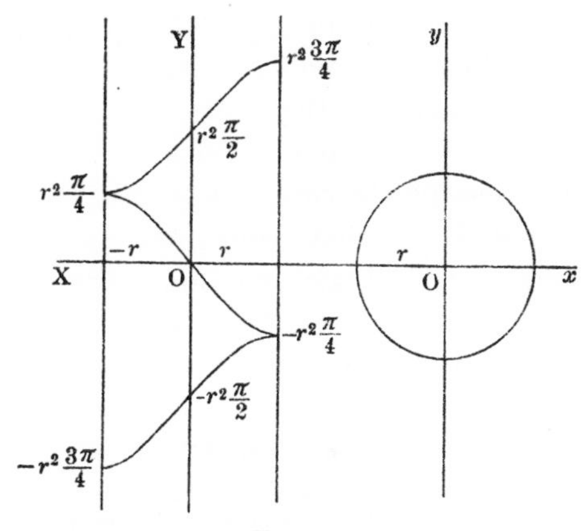

FIG. 16.

The apparatus should be so constructed then that the tracing point shall move parallel to the variable direction of this hypotenuse, while the guiding point describes the differential curve. This is effected by connecting the tracing point with a sharp-edged roller whose plane is vertical and moves so as to be always parallel to this hypotenuse. A weight presses this roller firmly upon the paper so that its point of contact can advance only in the plane of the roller.

The practical object of the integraph is the approximate evaluation of definite integrals; for us its application to the construction of π is of especial interest.

3. Take for differential curve the circle

$$x^2 + y^2 = r^2;$$

the integral curve is then

$$Y = \int \sqrt{r^2 - x^2}\,dx = \frac{r^2}{2}\sin^{-1}\frac{x}{r} + \frac{x}{2}\sqrt{r^2 - x^2}.$$

This curve consists of a series of congruent branches. The points where it meets the axis of Y have for ordinates

$$0, \quad \pm\frac{r^2\pi}{2}, \quad \cdots$$

Upon the lines $X = \pm r$ the intersections have for ordinates

$$r^2\frac{\pi}{4}, \quad r^2\frac{3\pi}{4}, \quad \cdots$$

If we make $r = 1$, the ordinates of these intersections will determine the number π or its multiples.

It is worthy of notice that our apparatus enables us to trace the curve not in a tedious and inaccurate manner, but with ease and sharpness, especially if we use a tracing pen instead of a pencil.

Thus we have an actual constructive quadrature of the circle along the lines laid down by the ancients, for our curve is only a modification of the quadratrix considered by them.

NOTES FOR
PART I AND PART II

NOTES

PART I — CHAPTER III

Gaussian Polygons. Up to the time of Gauss, no one suspected that it was possible to construct, with ruler and compasses, regular polygons other than those the number of whose sides could be expressed in one of the forms: 2^n, $2^n \cdot 3$, $2^n \cdot 5$, $2^n \cdot 15$. All of these were known to the Greeks. But Gauss proved as early as 1801[1] that whenever a prime number F_μ could be expressed in the form $2^{2^\mu}+1$, the construction of a regular polygon with F_μ sides was possible by Euclidean methods. It was then apparent that regular polygons not included in the Euclidean series, namely 17, 257, 65537, . . . sides, could be constructed under the same imposed conditions. And indeed Gauss's discussion led to the result[2], that the *only* regular polygons which it is possible to construct with ruler and compasses, are those the number P of whose sides can be expressed in the form

$$2^\alpha \cdot (2^{2^{\alpha_1}} + 1) \cdot (2^{2^{\alpha_2}} + 1) \cdot (2^{2^{\alpha_3}} + 1) \cdots (2^{2^{\alpha_s}} + 1),$$

where α . . . α_s are distinct positive integers and each $2^{2^{\alpha_i}}+1$ is a prime. The number of such polygons is small in comparison with the number of regular polygons which can not be constructed with the means employed. As Dickson has pointed out[3] the number of P's up to 100 is 24; up to 300 is 37 (all noted by Gauss); up to 1000 is 52; up to 1000000 only 206. Kraitchik has remarked[4] that there are only 30 polygons with an odd number of sides that are known to be constructible with ruler and compasses. These polygons have the following number of sides: 5, 15, 17, 51, 85, 255, 257, 771, 1285, 3855, 4369,13107, 21845, 65535, 65537, 196611, 327685, 983055, 1114129, 3342387, 5570645, 16711935, 16843009, 50529027, 84215045, 252645135, 286331153, 858993459, 1431655765,

---

[1] *Disquistiones arithmeticæ*, Leipzig, 1801, p. 664; *Werke*, v. 1., 2. Abdruck, 1870, p. 462; French ed. *Recherches Arithmétiques*, Paris, 1807, p. 488; Ger. ed. by Maser, Berlin, 1889, p. 447.

[2] This result was, in effect, stated, but not proved, by Gauss.

[3] L. E. Dickson, "On the number of inscriptible regular polygons", *Bull. N. Y. Math. Soc.*, Feb., 1894, v. 3, p. 123.

[4] Kraitchik, *Recherches sur la théorie des nombres*, Paris, 1924, p. 270.

4294967295. This set of numbers, together with 1 and 3, coincides with the divisors of $2^{32} - 1 = 1 \cdot 3 \cdot 5 \cdot 17 \cdot 257 \cdot 65537$.

The determination of the number of regular polygons which can be constructed for P less than a given integer is, then, bound up in the determination of the prime numbers F_μ. Now for only 18 values of μ has it been shown whether F_μ is prime or not, namely for the values of μ from 0 to 9 inclusive, and for 11, 12, 15, 18, 23, 36, 38, 73. In the first five of these cases, and in these alone, is F_μ prime. These five cases were noted by Fermat in the seventeenth century. It may well turn out that F_μ is not prime, for $\mu > 4$, although Eisenstein proposed as a problem[1]: "There are an infinity of prime numbers of the form $2^{2^\mu} + 1$".

The results already established in this connection may be set forth in tabular form[2]:

μ	Prime Factors of $F_\mu = 2^{2^\mu} + 1$	Discoverer	Year of Discovery
0-4	3, 5, 17, 257, 65537	Fermat	1640
5	$2^7 \cdot 5 + 1 = 641$ $2^7 \cdot 52347 + 1 = 6700417$	L. Euler	1732
6	Unknown but composite.	Lucas	1878
	$2^8 \cdot 9 \cdot 7 \cdot 17 + 1 = 274177$	Landry	1880
	$2^8 \cdot 5 \cdot 52562829149 + 1 + 67280421310721$	Landry and Le Lasseur	1880
7	Unknown but composite.	A. E Western, J. C, Morehead	1905
8	Unknown but composite.	A. E. Western J. C. Morehead	1909
9	$2^{16} \cdot 37 + 1 = 2424833$	A. E. Western	1903
11	$2^{13} \cdot 3 \cdot 13 + 1 = 319489$ $2^{13} \cdot 7 \cdot 17 + 1 = 974849$	A, Cunningham	1899
12	$2^{14} \cdot 7 + 1 = 114689$	E. A. Lucas and P. Pervouchine	1877
	$2^{16} \cdot 397 + 1 = 26017793$ $2^{16} \cdot 7 \cdot 139 + 1 = 63766529$	A. E. Western	1903
15	$2^{21} \cdot 579 + 1 = 1214251009$	M. Kraitchik	1925
18	$2^{20} \cdot 13 + 1 = 13631489$	A. E. Western	1903
23	$2^{25} \cdot 5 + 1 = 167772161$	P. Pervouchine	1878
36	$2^{39} \cdot 5 + 1 = 2748779069441$	Seelhoff	1886
38	$2^{41} \cdot 3 + 1 = 6597069766657$	J. Cullen, A. Cunningham, A. E. and F. J. Western	1903
73	$2^{75} \cdot 5 + 1 = 188894559314785808547841$	J. C. Morehead	1906

[1] G. Eisenstein, "Aufgaben", Crelle's *Journal*, v. 27, 1844, p. 87.

[2] The sources for the different results, except those of Fermat, are as follows, for the 13 different values of μ:

The labor expended in deriving these results has been enormous; to the layman who knows nothing of congruences in the theory of numbers, the facts found must seem almost to border on the miraculous. For, even when $\mu = 10$, a case not yet solved, F_μ contains 309 digits; but when $\mu = 36$, F_μ is a number of more than twenty trillion digits. Concerning it Lucas remarked[1] "la bande de papier qui le contiendrait ferait le tour de la Terre". For $\mu = 73$, Ball states that the digits in F_μ "are so numerous that, if the number were printed in full with the type and number of pages used in this book [*Mathematical Recreations*, fifth edition, 1911, 508 pages], many more

5. L. Euler, *Commentarii Academiœ Scientiarium Petrop.*, v. 6 (1732—3), 1738, p. 104; laid before the Academy of St. Petersburg, 26. Sept. 1732.
In his autobiography (Springfield, Mass., 1833, p. 38) the American calculator Zera Colburn records that while on exhibition in London, at the age of 8, he found "by the mere operation of his mind" the factors 641 and 6, 700, 417 of 4, 294, 967, 297 ($= 2^{32} + 1$). *Cf.* F. D. Mitchell, "Mathematical prodigies", *Amer. Journal of Psychology*, v. 18, 1907, p. 65.

6. Lucas, *Comptes Rendus de l'Académie des Sciences*, Paris, v. 85, 1878, p. 138; *Amer. Jour. Math.*, v. 1, 1878, p. 238; *Recreations mathématiques*, v. 2 (2e éd., 1896), p. 234—5. Landry, *Nouv. Corresp. Math.*, v. 6, 1880, p. 417.

7. Independent discoverers: Western, *Proc. Lond. Math. Soc.*, s. 2, v. 3, p. xxi—xxii. Abstract of paper read, April 13, 1905; Morehead, *Bull. Amer. Math. Soc.*, v. 11, p. 543—545, abstract of paper read April 29, 1905.

8. Western and Morehead, *Bull. Amer. Math. Soc.*, v. 16, 1909, p. 1—6; "each doing half of the whole work".

9, 12 (Western), 13, 16. *Proc. Lond. Math. Soc.*, s. 2, v. 1, 1903, p. 175; abstract of paper read May 14, 1903.

11. A. Cunningham, *Brit. Assoc. Rept.*, 1899, p. 653—4; the factors are here given as 319489 and 974489. The second number is incorrect, 4 and 8 being interchanged. The other forms of the correct factors were given by A. Cunningham and A. E. Western in *Proc. Lond. Math. Soc.*, s. 2, v. 1, 1903, p. 175. It is here noted also that there are no more factors of $F_\mu < 10^6$, and no other factor of $F_\mu < 10^8$, (μ not less than 14).

12, 23. E. Lucas, *Atti Accad. Torino*, v. 13 (1877—8), p. 271 [27 Jan., 1878]. *Mélanges math. ast. acad. Pétersb.*, v. 5, part 5, 1879, p. 505, 519, or *Bull. Acad. Pétersb.*, s. 3, v. 24, 1878, p. 559; s. 3, v. 25, 1879, p. 63; communication of results, for $\mu = 12$ and 23, found by J. Pervouchine, in Nov. 1877 and Jan. 1878. He notes that the integer $2^{2^{23}} + 1$ contains 2525223 digits.

15. M. Kraitchik, *Comptes Rendus de l'Académie des Sciences*, Paris, v. 180, p. 800, March, 1925; also *Sphinx-Oedipe*, v. 20, p. 24.

36. P. Seelhoff, *Zeitschrift math. u. Phys.*, v. 31, 1886, p. 174.

73. J. C. Morehead, *Bull. Amer. Math. Soc.*, v. 12, 1906, p. 449—451.

[1] E. Lucas, *Théorie des nombres*, Paris, v. 1, 1891, p. 51.

volumes would be required than are contained in all the public libraries of the world".

In not less than seven places[1], during the years 1640–58, did Fermat refer to $F_\mu = 2^{2^\mu} + 1$ as representing a series of prime numbers; but in no place did he claim that F_μ was always prime.

Gauss's Statement of his Polygon Results. In two passages the implication to be drawn from what Klein has written is, that Gauss published a proof that a regular polygon of p sides can not be constructed by ruler and compasses if p is a prime not of the form $2^k + 1$. The passages to which I refer are (pages 2, 16): (1) "Gauss added other cases [to Euclid's] by showing the possibility of the division into parts where p is a prime number of the form $p = 2^{2^\mu} + 1$, and the impossibility for all other numbers"; (2) "Gauss extended this series of numbers [Euclid's] by showing that the division is possible for every prime number of the form $p = 2^{2^\mu} + 1$ but impossible for all other prime numbers and their powers". Now the implication referred to above is not correct, as Pierpont interestingly set forth in his paper "On an undemonstrated theorem of the *Disquisitiones Arithmeticae*"[2]. That is, Gauss *did not give a proof* of the "impossibility" referred to in the quotations. But after proving the "possibility" described above he continued as follows:

"As often as p—1 contains other prime factors besides 2, we arrive at higher equations[3], namely, to one or more cubic equationas, if 3 enters

[1] Letter dated Aug. [?] 1640 to Frenicle (*Oeuvres de Fermat*, v. 2, 1894, p. 206); letter dated 18 Oct., 1640, to Frenicle (*Oeuvres*, v. 2, 1894, p. 208); *Varia Opera*, Toulouse, 1679, p. 162; Brassine's *Précis*, Toulouse, 1853, p. 142—3); letter dated 25 Dec., 1640, to Mersenne (*Oeuvres*, v. 2, p. 212—213); "De solutione problematum geometriconum per curvas simplicissimas et unicuique problematum generi proprie convenientes, Dissertatio tripartita" (*Oeuvres de Fermat*, v. 1, 1891, p. 130—131; French translation, v. 3, 1896, p. 120; *Varia Opera*, 1679 [reprint, 1861], p. 115); letter dated 29 August, 1654, to Pascal (*Oeuvres de Pascal*, v. 4, Paris, 1819, p. 384; *Oeuvres de Fermat*, v. 2, 1894, p. 309—310); letter to Sir Kenelm Digby, sent by Digby to Wallis, 19 June, 1658 (*Oeuvres de Fermat*, v. 2, 1894, p. 402, 404—5; French translation of the Latin, v. 3, 1896, p. 314, 316); letter dated August, 1659 to Carcavi, copy sent by Carcavi to Huygens 14 August, 1659 (*Corresp.* de Huygens no. 651; *Oeuvres de Fermat*, v. 2, p. 433—434).

[2] *Bull. Amer. Math. Soc.*, v. 2, 1895, p. 77—83.

[3] In his earlier discussion of an inscribed polygon of p sides, Gauss considers the equation $x^p - 1 = 0$ and the resulting equation got by dividing out the factor $x - 1$, where p is a prime.

once or oftener as a factor of $p-1$, to equations of 5th degree if $p-1$ is divisible by 5, etc. And we can prove with all rigour that these equations cannot be avoided or made to depend upon equations of lower degree; and although the limits of this work do not permit us to give the demonstration here, we still thought it necessary to signal this fact in order that one should not seek to construct other polygons than those given by our theory, as, for example, polygons of 7, 11, 13, 19 sides, and so employ one's time in vain."

Fermat's Theorem. This theorem (p. 17) was indicated by Fermat in a letter, dated 18 October 1640, to B. Frenicle de Bessy (*Oeuvres de Fermat*, v. 2, 1894, p. 209). Euler gave two proofs (*Comment. Acad. Petrop.*, v. 8 for 1736, 1741, p. 141, and *Comment. Nov. Acad. Petrop.*, v. 7 for 1758-59, 1761, p. 49). Other proofs are due to Lagrange (*Nouv. Mém. de l'Acad. de Berlin*, 1771) and to Gauss (*Disquisitiones Arithmeticæ*, § 49)

PART I — CHAPTER IV

Geometrical Constructions of the Regular Heptadecagon. The remark of Klein (p. 24, 32) that we posses as yet no method of construction of the regular polygon of seventeen sides, based upon considerations purely geometrical, is a little curious, since several constructions of this kind have been given. One by Erchinger was indeed reported by Gauss in 1825[1]. The construction is as follows:

Let D, B, G, A, I, F, C, E be points on a line determined by constructions about to be given. Let AB be a line of any length. Produce it both ways to C and D so that,

D B G A I F C E

$$AC \times BC = AB \times BD = 4\,\overline{AB}^2.$$

[1] *Göttingische gelehrte Anzeigen*, Dec. 19, 1825, no. 203, p. 2025; *Werke*, v. 2, p. 186—7. To Art. 365 of the *Disquisitiones Arithmeticae* Gauss added this note in his handwriting: "Circulum in 17 partes divisibilem esse geometrice, deteximus 1796 Mart. 30". *Cf. Werke* v. 1, p. 476 and v. 10_1, 1917, p. 3—4, 120—126, 488. The discovery of the result was first announced in the *Intelligenzblatt* of the *Allgemeine Literatur-Zeitung*, no. 66, 1 June, 1796, col. 554.

Further determine the points, E, G, on both sides of CA produced so that,

$$AE \times EC = AG \times CG = \overline{AB}^2;$$

and find the point F on the side A of the line BA produced, such that

$$AF \times DF = \overline{AB}^2.$$

Finally divide AE in I so that

$$AI \times EI = AB \times AF,$$

where AI is the smaller, and EI the larger part of AE. Then construct a triangle, in which each of two sides equals AB, the third being equal to AI. About this triangle describe a circle; then AI will be one side of the regular inscribed polygon of seventeen sides.

Gauss particularly remarks that the author gave a purely synthetic proof of this construction.

Another synthetic construction and proof dated "Dublin, 17th October, 1819" was published by Samuel James in the *Transactions of the Irish Academy*[1]. Yet another construction was given by John Lowry in *The Mathematical Repository*[2] for 1819. But the earliest published geometrical construction was given by Huguenin in his *Mathematische Beiträge zur weiteren Ausbildung angehender Geometer*, Königsberg, 1803, p. 283.

A score of geometrical constructions are assembled in A. Goldenring, *Die elementargeometrischen Konstruktionen des regelmässigen Siebzehnecks*, Leipzig, 1915. See also the review of this work in *Bull. Amer. Math. Soc.*, v. 22, 1916, p. 239—246, and my note "Gauss and the regular polygon of seventeen sides" in *Amer. Math. Monthly*, v. 27, 1920, p. 323—326.

The discovery that the regular polygon of seventeen sides could be constructed with ruler and compasses was not only one of which Gauss was vastly proud throughout his life, but also, according to Sartorius von Waltershausen[3], the one which decided him to dedicate his life to the study of mathematics. Archimedes expressed the wish that a sphere inscribed in a cylinder be inscribed on his tomb, as Ludolf van Ceulen did in connection with the value of π to 35 decimal

---

[1] V. 13 (1818), p. 175—187; paper read Jan. 24, 1820.

[2] N. s., v. 4, p. 160. Lowry's proof occupies p. 160—168.

[3] *Gauss zum Gedächtniss*, Leipzig, 1856, p. 16.

places, and Jacques Bernoulli with reference to the logarithmic spiral. So also, according to Weber[1], Gauss requested that the regular polygon of seventeen sides should be engraved on his tombstone. While this request was not granted, as it was in each of the other cases mentioned, it is engraved on the side of a monument to Gauss in Braunschweig, his birthplace.

Constructions in general with Ruler and Compasses. Regarding constructions as effected when intersections of circles with circles or lines, or of lines with lines may be determined, it can be shown that: *Every problem solved with ruler and compasses can be solved with compasses alone.* This was first shown by Georg Mohr in his *Euclides Danicus* published at Amsterdam in 1672; this work was reprinted in 1928 by the Danish Society of Sciences. Klein refers (p. 33) only to Mascheroni's proof of this result 125 years later, in his *Geometria del Compasso.* Of this work there were two French editions *Géométrie du Compas,* Paris, 1798 and 1828. From the first of these a German edition *L. Mascheroni's Gebrauch des Zirkels,* Berlin, 1825, was prepared by J. P. Gruson. The subject is treated in English by: A. Cayley, *Messenger of Math.*, v. 14, 1885, p.179—181; *Collected Papers,* v. 12, p. 314—317; by E. W. Hobson, in a presidential address, *Mathematical Gazette,* v. 7, 1913, p. 49—54; by H. P. Hudson, *Ruler & Compasses,* London, 1916, p. 131—143; and by J. Coolidge, *Treatise on the Circle and Sphere,* Oxford, 1916, p. 186—188.

Klein has noted (p. 33—34) that Poncelet first conceived the result that *given a circle and its center, every solution of a problem with ruler and compasses can be carried through with ruler alone.* A little later Klein states (p. 34) "we will show *how with the straight edge and one fixed circle we can solve every quadratic equation*". This is not possible; Klein should have had "with its center" after "one fixed circle". That the center be also given is very essential when only one circle is given. Hilbert suggested the problem: How many given circles in a plane are necessary in order to determine with ruler alone, the center of one of them? In 1912 D. Cauer[2] showed: (a) If two circles do not intersect in real points it is generally impossible to determine the center of either circle with ruler alone; (b) A center

---

[1] *Encyclopädie der elementaren Algebra und Analysis* bearbeitet von H. Weber. 2. ed. Leipzig, 1906, p. 362.

[2] *Mathematische Annalen,* v. 73, 1912, p. 90—94; v. 74, 1913, p. 462—464.

may be determined if the circles cut in real points, touch, or are concentric. About the same time J. Grossmann discovered a result which proved that *Every problem solvable with ruler and compasses can also be solved with ruler alone if we are given, in the plane of construction, three linearly independent circles.* Correct proofs of this result were given by Schur and Mierendorff.

From this it is clear that every construction with ruler and compasses can be effected with a ruler, and compasses with a fixed opening. Constructions of this kind were found already in the tenth century by Abû'l Wefâ of Bagdad[1]. With such means, in the sixteenth century, certain problems of Euclid were solved by Cardano, Ferraro, and Tartaglia. At Venice in 1553 G. B. Benedetti published a little treatise, *Resolutio omnium Euclidis problematum, aliorumque ad hoc necessario inventorum, una tantumodo circuli data apertura.* In English the topic is treated in a rare pamphlet translated from the Dutch by Joseph Moxon[2], and in an article by J. S. Mackay[3].

Every problem whose solution is possible by ruler and compasses can be also solved with a two edged ruler alone, whether the edges are parallel or meet in a point. For some of the literature in this connection the following sources may be consulted: *Nouvelle Corresp. Math.*, v. 3, 1877, p. 204—208; v. 5, 1879, p. 439—442; v. 6, 1880, p. 34—35; Akademie der Wissen., Vienna, *Sitzungsberichte*, Abt. IIa, v. 99, 1890, p. 854—858; *Bolletino di Matematiche e di Scienze fisiche e naturali*, v. 2, 1900—01, p. 129—145, 225—237.

PART II — CHAPTER II

Irrationality of π. Klein wrote (p. 59): "After 1770 critical rigour gradually began to resume its rightful place. In this year appeared the work of Lambert: *Vorläufige Kenntnisse für die, so die Quadratur*

[1] "Woepcke" Analyse et extrait d'un recueil de constructions géométriques par Aboûl Wafâ", *Journal Asiatique*, 1855.

[2] *Compendium Euclidis Curiosi: or, geometrical operations. Showing how with a single opening of the Compasses and a straight ruler all the propositions of Euclid's first five books are performed.* London, 1677. Moxon does not tell us who the author of the Dutch treatise was.

[3] "Solutions of Euclid's problems, with a ruler and one fixed aperture of the compasses, by the Italian geometers of the sixteenth century", *Proc. Edinb. Math. Soc.*, v. 5, 1887, p. 2—22.

. . . *des Cirkuls suchen.* Among other matters the irrationality of π is discussed. In 1794 Legendre in his *Éléments de Géométrie* showed conclusively that π and π^2 are irrational numbers." The implication of this note is that Lambert did not discuss the irrationality of π conclusively and that Legendre did. How both of these points of view are essentially incorrect will appear in what follows. Klein was simply reproducing the erroneous statements of Rudio[1]; but after Pringsheim's careful study in 1898[2], Lambert's proof emerged as "ausserordentlich scharfsinnig und im wesentlichen vollkommen einwandfrei", while Legendre's remained "in Bezug auf Strenge hinter Lambert weit zurück".

As in the later proof of the transcendence of π, so here when its irrationality was in question, discussion of e is fundamental. The irrationality of e and e^2 was shown, substantially, by Euler in 1737[3] and he gave the expression for e as a continued fraction on which Lambert's proofs of the irrationality of e^x, tan x and π rest. Starting with Euler's development[4]

$$\frac{e-1}{2} = \frac{1}{1+}\,\frac{1}{6+}\,\frac{1}{10+}\,\frac{1}{14+}\,\frac{1}{18+}\,\text{etc.},$$

Lambert found

$$\frac{ex-1}{ex+1} = \frac{1}{2/x+}\,\frac{1}{6/x+}\,\frac{1}{10/x+}\,\frac{1}{14/x+}\,\textit{etc.},$$

and since

---

[1] F. Rudio: *Archimedes, Huygens, Lambert, Legendre, vier Abhandlungen über die Kreismessung.* Leipzig, 1892, p. 56f. This error is also reproduced by B. Calò in Enriques's *Fragen der Elementargeometrie,* II. Teil, 1907, p. 315; by D. E. Smith in Young's *Monographs on Topics of Modern Mathematics,* 1911, p. 401. The matter was correctly set forth by T. Vahlen in *Konstruktionen und Approximationen,* Leipzig, 1911, p. 319f.

[2] A. Pringsheim: "Über die ersten Beweise der Irrationalität von e und π", Bayerische Akad. der Wissen., *Sitzungsberichte,* mathem.-phys. Cl., v. 28, 1899, p. 325–337.

[3] "De fractionibus continuis", *Comment. acad. de Petrop,* v. 9, 1744, p. 108. Presented to St. Petersburg Academy, March. 1737.

[4] L. Euler: *Introductio in analysin infinitorum.* Tomus Primus, Lausannae, 1748, p. 319. This work was finished in 1745; *Cf.* G. Eneström, *Verzeichnis* etc., Erste Lieferung, p. 25.

$$\frac{e^x - 1}{e^x + 1} = \frac{e^{x/2} - e^{-x/2}}{e^{x/2} + e^{-x/2}} = \tanh \frac{x}{2} = \frac{1}{i} \tan \frac{ix}{2}, \text{ if } z = \frac{ix}{2},$$

$$\tan z = \frac{1}{1/z -} \frac{1}{3/z -} \frac{1}{5/z -} \frac{1}{7/z -} \frac{1}{9/z -} \cdots.$$

He then proved the theorems:

1. *If x is a rational number different from zero, e^x can never be rational.*

For $x = 1$, we have as special case the irrationality of e.

2. *If z is a rational number different from zero, tan z can never be rational.*

For $z = \pi/4$, $\tan \pi/4 = 1$, and hence as a special case the irrationality of π.

The part of Lambert's *Vorläufige Kenntnisse* to which Klein refers contains some formulae without proof, and no analytical developments, and was rather intended to serve as a popular survey of the treatment of the topic. With it must be considered the scientifically remarkable "*Mémoire*" of 1767[1]. Here "mit minutiöser Genauigkeit" Lambert proves the convergence of the expression for tan z as a continued fraction. Pringsheim dwells on the "astounding" nature of these considerations at this period in the history of mathematical thought. For of such considerations Legendre was innocent, as well as the great Gauss in his 1812 memoir on hypergeometric series, and others, till a much later period.

"Thus the Lambert memoir contains the *first*, and for many years, the *only* example of what we now consider really rigorous developments of functions as converging continued fractions, in particular, that for tan z given above."

Measurement of a Circle. By considering inscribed and circumscribed polygons up to 96 sides Archimedes arrived at the result that the ratio of the circumference of a circle to its diameter is less than $3\frac{10}{70}$ but greater than $3\frac{10}{71}$. The following table exhibits the perimeters of regular inscribed and circumscribed polygons of a circle with a unit diameter (Chauvenet, *Treatise on Elementary Geometry*, Philadelphia, 1870, p. 161).

---

[1] "Mémoire sur quelques propriétés remarquables des quantités transcendantes circulaires et logarithmiques". Lu en 1767. Printed in 1768 in *Hist. de l'acad. royale des sciences et belles-lettres*, Berlin, Année 1761 (!), p. 265—322.

Number of sides	Perimeter of circumscribed polygon	Perimeter of inscribed polygon
4	4.0000000	2.8284271
8	3.3137085	3.0614675
16	3.1825979	3.1214452
32	3.1517249	3.1365485
64	3.1441184	3.1403312
128	3.1422236	3.1412773
256	3.1417504	3.1415138
512	3.1416321	3.1415729
1024	3.1416025	3.1415877
2048	3.1415951	3.1415914
4096	3.1415933	3,1415923
8192	3.1415928	3.1415926

The remarkable approximation 355/113 for π is correct to six places of decimals. It seems to have been first given by a Chinese, Tsu Ch'ung-ching (5th century), and later by Valentin Otho (16th century) and Adriaen Anthonisz (17th century). Grunert gave a geometrical construction for π based on the fact that $355/113 = 3 + 4^2/(7^2 + 8^2)$, *Archiv der Mathematik und Physik*, v. 12, 1849, p. 98. Another construction was given by Ramanujan in *Journ. Indian Math. Soc.*, v. 5, 1913, p. 132 (also in *Collected Papers of Srinivasa Ramanujan*, Cambridge, 1927, p. 22, 35).

Euler's Formula. The formula

$$e^{ix} = \cos x + i \sin x$$

was first given by Euler in *Miscellanea Berolinensia*, v. 7, 1743, p. 179 (paper read 6 Sept. 1742), and again in his *Introductio in Analysin*, Lausanne, 1748, v. 1, p. 104. He gave also

$$e^{-ix} = \cos x - i \sin x.$$

The equivalent of the form

$$ix = \log (\cos x + i \sin x$$

was given earlier by Roger Cotes (*Philosophical Transactions*, 1714, v. 29, 1717, p. 32) as: "Si quadrantis circuli quilibet arcus, radio

CE descriptus, sinum habeat CX, sinumque complementi ad quadrantem XE: sumendo radium CE, pro Modulo, arcus erit rationis inter $EX + XC\sqrt{-1}$ & CE mensura ducta in $\sqrt{-1}$." See also Cotes, *Harmonia Mensurarum*, 1722, p. 28.

PART II — CHAPTER IV

In the course of the discussion on pages 61—74 it is assumed that there are an infinite number of prime numbers. One of the neatest proofs of this fact was given by Euclid (about 300 B.C.) in proposition 20, book 9 of his Elements.

On page 77, in considering the relation $y = e^x$, Klein made a slight slip when he wrote: "To an algebraic value of x corresponds a transcendental value of y, and conversely." "Conversely" leads to the statement, to a transcendental value of y corresponds an algebraic value of x. But a proof of this has nowhere been given; indeed the result is not true, in general. To correct delete "conversely" and add: "To an algebraic value of y corresponds a transcendental value of x."

A series of hardcover reprints of major works in mathematics, science and engineering.
All editions are 5⅜ × 8½ unless otherwise noted.

Mathematics

Theory of Approximation, N. I Achieser. Unabridged republication of the 1956 edition. 320pp. 49543-4

The Origins of the Infinitesimal Calculus, Margaret E. Baron. Unabridged republication of the 1969 edition. 320pp. 49544-2

A Treatise on the Calculus of Finite Differences, George Boole. Unabridged republication of the 2nd and last revised edition. 352pp. 49523-X

Space and Time, Emile Borel. Unabridged republication of the 1926 edition. 15 figures. 256pp. 49545-0

An Elementary Treatise on Fourier's Series, William Elwood Byerly. Unabridged republication of the 1893 edition. 304pp. 49546-9

Substance and Function & Einstein's Theory of Relativity, Ernst Cassirer. Unabridged republication of the 1923 double volume. 480pp. 49547-7

A History of Geometrical Methods, Julian Lowell Coolidge. Unabridged republication of the 1940 first edition. 13 figures. 480pp. 49524-8

Linear Groups with an Exposition of Galois Field Theory, Leonard Eugene Dickson. Unabridged republication of the 1901 edition. 336pp. 49548-5

Continuous Groups of Transformations, Luther Pfahler Eisenhart. Unabridged republication of the 1933 first edition. 320pp. 49525-6

Transcendental and Algebraic Numbers, A. O. Gelfond. Unabridged republication of the 1960 edition. 208pp. 49526-4

Lectures on Cauchy's Problem in Linear Partial Differential Equations, Jacques Hadamard. Unabridged reprint of the 1923 edition. 320pp. 49549-3

The Theory of Branching Processes, Theodore E. Harris. Unabridged, corrected republication of the 1963 edition. xiv+230pp. 49508-6

The Continuum, Edward V. Huntington. Unabridged republication of the 1917 edition. 4 figures. 96pp. 49550-7

Lectures on Ordinary Differential Equations, Witold Hurewicz. Unabridged republication of the 1958 edition. xvii+122pp. 49510-8

Mathematical Methods and Theory in Games, Programming, and Economics: Two Volumes Bound as One, Samuel Karlin. Unabridged republication of the 1959 edition. 848pp. 49527-2

Famous Problems of Elementary Geometry, Felix Klein. Unabridged reprint of the 1930 second edition, revised and enlarged. 112pp. 49551-5

Lectures on the Icosahedron, Felix Klein. Unabridged republication of the 2nd revised edition, 1913. 304pp. 49528-0

On Riemann's Theory of Algebraic Functions, Felix Klein. Unabridged republication of the 1893 edition. 43 figures. 96pp. 49552-3

A Treatise on the Theory of Determinants, Thomas Muir. Unabridged republication of the revised 1933 edition. 784pp. 49553-1

A Survey of Minimal Surfaces, Robert Osserman. Corrected and enlarged republication of the work first published in 1969. 224pp. 49514-0

The Variational Theory of Geodesics, M. M. Postnikov. Unabridged republication of the 1967 edition. 208pp. 49529-9